Digital Urban Acupuncture

Salvatore Iaconesi · Oriana Persico

Digital Urban Acupuncture

Human Ecosystems and the Life of Cities in the Age of Communication, Information and Knowledge

Springer

Salvatore Iaconesi
ISIA Design Florence
Rome
Italy

Oriana Persico
ISIA Design Florence
Rome
Italy

ISBN 978-3-319-43402-5 ISBN 978-3-319-43403-2 (eBook)
DOI 10.1007/978-3-319-43403-2

Library of Congress Control Number: 2016948118

Printed on acid-free paper

This Springer imprint is published by Springer Nature
The registered company is Springer International Publishing AG Switzerland

Preface

In Alfred Hitchcock's Psycho, Norman Bates is split between two houses (Zizek 2006): the modern horizontal motel and his mother's gothic house. He is constantly running between the two, not being able to find his own. This parallels his complex personality, impersonations, and (perceived) identity(ies): The two architectures represent Norman's identities, and the continuous oscillations between the one and the other mimic their confrontation. Psycho is a product of modernity, in which this type of struggle is evident: In the change of the century, and after the wars, "home" becomes a complex concept, and a space of conflict, which becomes even more radicalized as the times of the revolutions of the young generations start appearing at the horizon, first with the signals of music and art, and then with the sexual revolution, the drugs, and the nomadic lifestyles which come with them at both physical, mental, and economic levels.

The postmodern dimension tends to obfuscate this dimension. If a postmodern architect was to be involved in the design of Psycho's location, the two buildings, with some probability, would have been collapsed together, in one of Ghenry's style mashups, where the poles of the antagonism could have compenetrated one into the other, combining them into a new hybrid entity. With all probability, in this case, Norman would have had no need to kill his victims, as the tension of running around between the two places (and identities) would have collapsed as well, into a third space of hybridity. At last, Norman Bateson would have been home.

Defining home also implies defining what is not home. And, thus, it implies the definitions, on the one hand, of public spaces and, on the other hand, of the private and intimate ones, which become ever more layered and fragmented: from a house, to the teen's room, to headphones with which to create and personalize our space, to the Internet, in which we can multiply ourselves and potentially create infinite numbers and modalities of public, private, and intimate contexts, identities, and environments.

The genesis of the Sony Walkman, for example, is perfectly fitting in describing this sort of (r)evolution.

As the technical opportunities allowed for smaller devices with higher quality to become ever closer to the body, architecture changed.

At first, you experienced music in the concert hall, with the selection of the musical landscape being performed administratively, by experts, curators, and politicians. Then, music moved into the houses of the rich, with instruments and expensive devices: They could invite, decide, and experience music in their own terms, the elites of culture. Then it started moving into every home, with record players and the radio: You could choose what to listen to, but there would still be fights over the radio channel to hear, or on what record to buy, a fight around consensus, authority, and the micropolitics of the house. And then, the music arrived behind the closed doors of teenagers' rooms, with all the signals of their desire for privacy, discouraging intrusion. Music also left the home, with the "ghetto blasters," the battery powered radios which you could carry on your shoulder and lay down, to broadcast music around yourself: Where your music was, the territory was your, nomadic, mobile, with you at the center. From the ghetto blaster the walkman arrived: By wearing headphones, you could completely personalize the space around you, connecting it to visions, memories, and desires. The walkman was probably the first available augmented reality consumer device: wearable augmented reality. This signed a radical transition: The compresence of multiple sounds in the same space, "attached" to the bodies of their owners, transformed urban locations into open source spaces of perceptions; it became possible to wear our own interpretation of space.

Multiple walkmans in the same space meant multiple interpretations of it: While someone was remembering romance, someone else could be experiencing violence, or excitement, or sadness, through the technological device connected to their body, extending it and its spatial capabilities.

Myriads of microhistories, side by side, compenetrating each other, paired to dress codes, behaviors, intentions, attitudes, and composed continuous flows of tactical usages of space: different music for taking the train, to wait in line, to walk in a certain neighbourhood or in another, for each person, simultaneously, in enormous numbers.

Architecture exploded: By listening to a different music, you could experience the same place in entirely different ways, also connecting to different times and places, with memories, past experiences, information (audio guides in your headphones), and emotions.

Then came the internet, further radicalizing this process, from both computational and relational points of view. Computational, for the action of algorithms, producing and presenting content and opportunities for interaction at every step, in ways which are mediated through the availability of network and connectivity, and by the action of the algorithms which select what you will experience at a certain space/time on your smartphone, and who you will connect with, what reviews you will see instead of others, what you will find in your search engine query, and what will other people see about what you expressed online. And relational, with people, object, places, a sort of novel set of superpowers which allow for almost telepathic connection with other identities, information producing or absorbing devices, remote vision through imaging, video, or tele-presence, and much more.

Time definitively became nonlinear, multiple, and emergent, and so did spaces, contexts, identities.

Norman Bateson could have easily been Norman, Psycho, and a dozen other personalities at the same time, with different Relational Ecosystems and interconnection with objects and locations, no matter where/when he was.

In all of this, the public, private, and intimate spaces are not only becoming progressively hybrid in these senses (as they integrate digital and physical, and as they become interconnective and nonlinear, instead of identitarian and linear), but also mediated. The readability and understandability of these spaces become more opaque and incomprehensible every day, and out of our possibilities for control, however individual or autonomous. Accessibility and openness are only at the surface, at the interface level, and we are all progressively loosing the ability to understand what lies below and about how to get big pictures in unmediated, autonomous ways.

In this scenario, this book constitutes a small step which could prove to be helpful to create interventions in both of these senses.

Its objective is to introduce a methodology—Digital Urban Acupuncture—which allows to gain better understandings about the Relational Ecosystems found in territories (and, particularly, in cities), between human and nonhumans, people and buildings, devices and plants, and so on, and to use these understandings to design—autonomously and collaboratively—interventions in the posthuman relational fabric of the city. This book is about a method using which it is possible to try to understand the city as an hyper-complex living being—a body, with energies, emotions, and behaviors, at micro-, meso-, and macrolevels—in which millions of identities interweave relations to punctually stimulate this body through performative actions—the needles—in order to create tangible, collaborative, and participatory effects.

To do this, a reflection is necessary on the possibility to reappropriate our public, private, and intimate spaces, and about how to perform across them as individuals, identities, communities, cultures, and societies.

This is what we dedicate this book to: not only to learn new methods and techniques, but also to reflect on how to work together in order to reclaim our spaces in socially constructive, autonomous, sustainable, solid, caring ways.

Rome, Italy

Salvatore Iaconesi
Oriana Persico

Reference

Zizek S (2006) The pervert's guide to cinema. Amoeba Film, UK

Acknowledgements

This book would have not been possible without the contribution of many people, organizations, and institutions. They are too many to list, and for this, we invite all readers to look throughout the Web sites and references listed in this book: There you will find people, institutions, universities, communities, and workgroups without whom we would have not been able to perform this research.

Here, we will highlight a few inspirations which have made this book possible, and some institutions and organizations which have always supported us.

Inspirations: Massimo Canevacci, who is maybe the one who introduced us to a possibilistic vision of the city; Antonio Caronia, who has opened up, for us, the opportunities which come from understanding how powers manifest themselves; Derrick de Kerckhove, who we do not see as often as we would like, and who has inspired us into letting understand the profound psychological implications of technologies; Marco Casagrande, whom we have never met in person yet and who has nonetheless had a great impact on the ways in which we think, him and the wonderful anarchist grandmothers which we have learned to recognize as one of the most interesting examples of how to bring true, lasting transformation in cities; Franco Avicolli, for having believed in two unknown artists an researchers with a vision, many years ago in Mexico City, and for all the inspiring conversations on the meaning of cities in our cultures; Luisa Valeriani, for all her help in making us understand the implications of being performers, instead of "only" artists and researchers; Franca Formenti, for being a relational, interconnective performer in everything she does, which is one of the secrets on how to transform communities; Alex Giordano, for being a living example of the possibilities which come when you let Kronos and Kairos come together, and for explaining the implications which come if you do this, through his life; Marco Stancati, for realizing the power of "augmenting" the experience of our cities through data, information, and knowledge; Luca De Biase, for his continuous, silent, and effective support and for focusing on the necessity for sense-making and participatory action in our times; Tiziana Terranova, for all the enormous effort spent on achieving a more just, balanced and inclusive world, for all the precious insights which she constantly provides, and for the wonderful walks in Naples; all the wonderful people at the University of Athens, Dimitris Charitos, Daphne Dragona, Iouliani Theona, and all the others, for being one of the most advanced observers of the evolution of our cities; Penny Travlou, for being our most trusted personal ethnographer and for constantly sincere participant observation, a rhizome in the network; Michel Bauwens, for his enormous energy and dedication to the possibilities and opportunities which come when you are able to imagine a new world; Christian Iaione, for being a constant supporter, advisor, enabler, and friend; Mark Cote, for being Italian, Canadian, Australian, and English at the same time and for all that this implies in terms of imagining a human-centered, effective, reflective research practice; Maurizio Napolitano, for his hard-core approach to the commons, which achieves the possibility to imagine an open world, with a smile, against all odds; Layne Hartsell, for being one of the most sincere people we know and one of the most profound and imaginative thinkers on the mutations of our times; Simone Arcagni, for being so energetic and for always amazing us with something we did not know; Alessandro Tartaglia, for his passion and determination, which is an example for us all, even when it is too much ☺.

Organizations and Institutions: the Eisenhower Fellowships, which has believed in us by enabling to start the Human Ecosystems through the support of their Consequential Outcomes program; the Yale World Fellowships and the TED Fellowships, for providing a supporting, engaged, caring environment in which people are the only thing worthwhile investing in; the City of Bologna, for having understood the importance of reclaiming our digital public spaces, right from the start; Universidade Metodista and SESC Brazil, for providing a world of opportunities and for sincerely understanding the importance of our work; transmediale festival, for believing in us and providing constant, critical support; Ars Electronica, for supporting us through sincere, dedicated efforts and participation; New Haven City Administration, for understanding the implications of our work immediately and as a caring, interested institution; the P2P Foundation, for all the constant support and the true relational environment which it creates; LabGov, for the incredibly insightful approach on the legal and societal implications which come from commoning and, in general, from imagining a collaborative society; SOS, the Open Source School, for being an incredible new form of school which was just waiting to exist; Nefula, for the continuous support and for the insightful contributions to the visions on the future of cities.

For everyone else: You know who you are and how much we love and respect you (and given all of the examples and references, you are probably inside the book already).

Contents

List of Figures

Chapter 1
Introduction

Abstract This chapter provides a general introduction about the research process which brought to the writing of the book, and then synthesizes the its contents describing them on a chapter-by-chapter basis.

This book represents a reflection on the current state of the research process which has led us through 4 years of explorations of the transforming landscapes of information, knowledge, communication, and relation in cities worldwide.

In these few years, things have radically changed.

Things that were impossible became possible. Practices that seemed strange, or even weird, became normal. Processes whose results appeared to be unlikely became likely, so much that now we are using them as our main inspiration about what the future of our cities could be.

This transformation has not been conflict-free.

Job markets have changed. People have protested. Relationships have assumed new forms. Power architectures have evolved. Distributions of wealth, health, well-being, happiness, and access to opportunities have changed, not always for the better.

There have been migrations, revolts, crimes, and unethical behaviors.

Most of the time, all of this change found its direct or indirect origin in the massive amount of data, information, and knowledge which has become available and accessible to multiple types of subjects—citizens, designers, entrepreneurs, public administrators, activists, artists, secret services, and more—through the wide presence of ubiquitous technologies.

Our cities have become filled with technologies, so much that any gesture, motion, traversal, and action of our daily lives generate data and information and can be used to generate knowledge and wisdom. Whether it is by flipping on a light switch; purchasing a tomato at the supermarket; having fun at a bar with a Wi-fi connection; cooking in our technological kitchen; working at our office while interacting with our friends and relatives online; driving our car; or just, simply, walking, in our cities full of sensors, cameras, devices, antennas, network connected objects and infrastructures, and other technologies: each of these actions generate data.

S. Iaconesi and O. Persico, *Digital Urban Acupuncture*
DOI 10.1007/978-3-319-43403-2_1

Whether we realize it or not, whether consciously or unconsciously, whether we are on social networks or not, and whether we possess some "smart" object in our pocket or in our home or not, we constantly generate data and information, in direct and indirect ways.

These data can be used for multiple purposes: to sell us consumer products through targeted advertising; to find patterns for fighting terrorism; to make new scientific discoveries which could create better scenarios for health, pollution, energy consumption, food production and usage, public transport, safety, development, and more.

Patterns (or their absence) can be found in data and information, generating knowledge. Patterns can be found in knowledge, generating wisdom.

In this book, we will focus on one thing which can be done with all of these data and information: understanding its flows—and the resulting flows of knowledge, wisdom, emotion, and, in general, communication—in cities, allowing us to gain insights about how to make interventions in it, creating a better, more inclusive, participatory, collaborative place.

We call this process "Digital Urban Acupuncture" (DUA), and through the years, we have developed a methodology for it, which can be useful for researchers, citizens, public administrators, urban planners, activists, entrepreneurs, designers, and artists.

In DUA, we take into consideration technologically rich environments (mainly cities, although rural environments are progressively becoming rich in ubiquitous technologies, so much that there are wonderful perspectives in applying DUA in these spaces), and we observe them through their attitude of spontaneously generating digital data and information under many forms: social networks; sensor networks; applications and services; real-time or off-line datasets; or through more creative ways which can include urban installations or interventions, citizen activation processes, gamification techniques, and more, as we will see throughout the book. In this process, we harvest large amounts of data and information, which we analyze using multiple techniques and combining multiple disciplines, from computational sociology, to digital ethnography, to network theories (e.g., Actor–Network Theory (ANT)).

Through this observation process, we aim at describing what we have defined the Relational Ecosystem: the descriptions and models according to which data, information, knowledge, and wisdom flow across (and through) the subjects of the city, whether they are individual citizens, organizations, companies, groups, "tribes," administrations, or else.

The Relational Ecosystem of the city can be seen as a landscape, superimposed on the physical landscape of the city, but with different behaviors: Its time frames are different and nonlinear; it moves, in fluid ways; it interconnects (and separates) regardless of the physical proximity; it is emergent, generative, and spontaneous, like an ubiquitous garden made of spontaneous vegetation which, instead of being carried on by winds, animals, and human beings, gets spread through digital networks, Wi-fi signals, mobile networks, interactions, transactions, and more.

The Relational Ecosystem can be interpreted by integrating the contributions coming from a range of other disciplines, such as anthropology, sociology, statistics, architecture, and arts and design.

By gaining insights about the Relational Ecosystem, cities—or parts of them, maybe in specific times or contexts—can be described as organisms, their functioning and behavior closely related—and, in some cases, determined—by these flows. Just as, to an extent, the life of a human body can be described in terms of the flows of blood, food, air, lymph, and electricity, exchanging substances, information, resources, stimuli, feedbacks, and more, between nodes as tiny as cells, as large as organs, and as diffused as the immune system.

This vision of the city as a complex system of flows gives rise to the possibility to try and interpret them and their interruptions, obstructions, blockages, interferences, deviations, jams, and dispersions, using techniques as varied as the ones derived from data and network sciences and the ones related to anthropology, psychology, and communication sciences. Through these understandings, it is possible to plan, design, and conduct interventions whose aim is to re-establish certain flows, to create new ones or to constitute the possibility for the emergence of novel ones in different, often unexpected, ways.

This process is the one we define as DUA: understanding the flows in the Relational Ecosystem of the city to design interventions to influence them; using design, architecture, policy-making, planning, restoration, and recontextualization processes for spaces and places, education processes, artistic intervention, and more; and also providing citizens with the tools and methodologies which can be used to design their own interventions.

Through DUA, this process can be done in informed, active, inclusive, participatory ways, producing positive interactions and also confronting with the value which is represented by the existence of differences, of conflicts, of transgressions, and of the social, spatial, political, and psychological configurations which they dynamically imply, through their presence, absence, fuzziness, preciseness, persistence, and volatility.

On top of that, DUA also allows for an interesting set of reflections on the times of the city. While its technological characteristics make it suitable for all the real-time analysis of the urban environment, DUA also has the qualities which allow analyses of longer time spans, highlighting the different speeds at which information, knowledge, innovation, and creativity flow within the city and across it.

In the following chapters, we will explore the notions (technical, technological, and methodological) which enable DUA and explore some notable use cases. The structure of the book is shown in visual form in Fig. 1.1.

Chapter 2 is a short chapter which describes how to use this book. The book is, in fact, composed of multiple harmoniously interweaved and interconnected experiences: The text, the images, the information visualizations, and the interviews are connected to interactive online experiences such as videos, discussion spaces, social network conversations, datasets, and maps. This chapter will introduce the reader to the visual diagrams which explain, in easy and accessible ways, how to

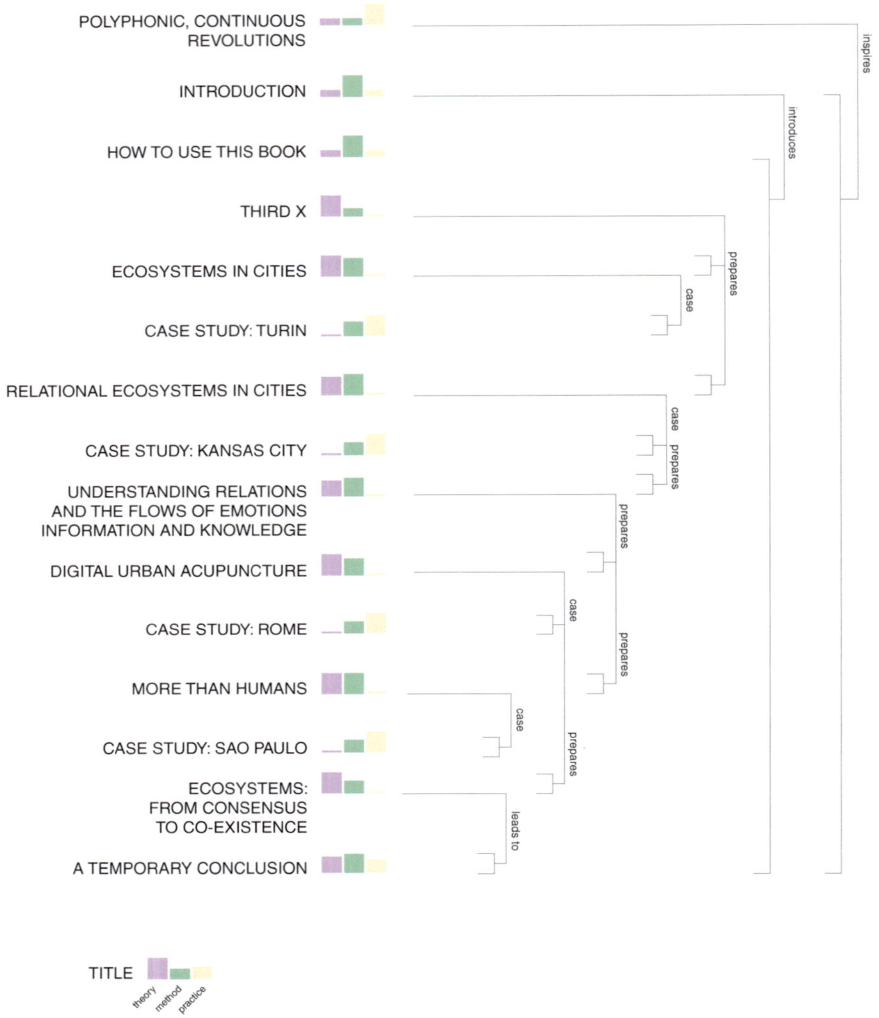

Fig. 1.1 The architecture of the book, showing how it covers the various parts of the DUA methodology

access all of its characteristics, from paper to the digital realm, and across devices (computers, smartphones, and tablets). This is commonly described as an "expanded book."

Chapter 3 introduces the concept of the Third Infoscape, which will be used throughout the book. Starting from Michel De Certeau's idea of the revolutionary potential of people's daily lives, and from the fact that people continuously reprogram the spaces they live in, we will explore the interconnection with several forms of emergence, of hybridization of spaces and contexts, and of *brassage* (remix), from Soja's and Bhabha's Third Space, to Clément's Third Landscape, to Casagrande's Third Generation City, to Pistoletto's Third Paradise. These, together, will allow us to conceive the new definition of the Third Infoscape, in which all of these concepts come together and integrate into a view of the world in which physical and digital domains are convergent through the presence and use of ubiquitous technologies and in which data and information constitute an emergent, relational part of the landscape.

In this chapter, an interview with Gilles Clément will show how these Thirds (landscape, space, city, infoscape, and paradise) can contribute to a more open, shared, imaginative society.

Chapter 4 will start off from the definition of the Third Infoscape and describe cities as ecosystems in which networks of interconnected human and non-human entities establish the flows of information, knowledge, emotion, and opinion. Several examples will be given for this concept, leading to the description of cities as a polyphony of expressions, of microhistories, tactics, and strategies, and about how they can be collected using data and information. The chapter will end with the exploration of the legibility of all of the micro- and macrohistories which unfold in territories and will describe their use according to the logic of coexistence.

At the end of the chapter, an interview with Carlo Ratti will introduce us to the potentialities and problems of taking into considerations large quantities of data and information for strategic tasks and in order to empower citizens to organize themselves and to collaborate.

Chapter 5 will present the first case study on the city of Turin, in Italy. This will address the observation of a full year (2011) in the city through social networks, focusing on the many languages which are spoken in the city, as evidence of the multiple cultures which inhabit it. Places, crossings, emotions, communication, opinions: the city is different according to the cultural eyes from which you look at it. Building upon Kevin Lynch's concepts of imageability and readability, the city shows its multiple lives and the ways in which they relate, interact, move, interconnect, and divide.

Chapter 6 will start from the city considered as an ecosystem and focus on its relational aspects, defining the Relational Ecosystem. The chapter will introduce the concept and move on to the description of how relations in the city can be discovered, observed, and measured, starting from the types of data and information which can be captured from the various types of actors in the ecosystems (quantitative, qualitative, classification using ontologies and folksonomies, topic and emotion discovery using natural language analysis, topic/emotion networks and their evolution in time, relational networks, and their evolution in time), and the types of relationships which can be described and observed in this perspective (strength and persistence of relations; vertical, horizontal, and transversal relations; liminal relations).

At the end of the chapter, an interview with Massimo Canevacci will frame the Relational Ecosystem in its anthropological implications, from the emergence of polyphonic performative actions, to dissent and conflict, to the possibility to observe and participate in the creation of imaginaries and actions.

Chapter 7 will feature another case study, through the observation of Kansas City's innovation ecosystem as expressed on social networks and through a series of institutional data sources. The case study will use the concepts defined in the previous chapter to highlight the architecture of the relations in the city's innovation ecosystem, its social, strategic, and political implications, and the understanding how to use these insights to perform meaningful, effective action in the city.

In Chapter 8, the notion of the Relational Ecosystem will be expanded, in order to understand what roles people and other entities may have in the ever-changing network of relations. The chapter introduces a methodology according to which the nodes of the Relational Ecosystem can be classified with respect to their characteristics and the ways in which it is possible to define further classification schemes to achieve different goals. Different classification schemes provide meaningful insights about the ways in which data, information, knowledge, emotion, and opinion flow in the city: information brokers, hubs, experts, influencers, amplifiers, bridges, and more. The focus here is, again, ecosystemic, in which human and non-human entities participate (organizations, for example, or objects, plants, and environments connected through sensors, or even data and information sources of various kinds which may be present in the ecosystem).

Chapter 9 will bring all of the previous elements together, defining a method for DUA. The Third Infoscape of the city gives rise to the Relational Ecosystem of the city through which it is possible to gain insights about the flows of data, information, knowledge, emotion, opinion, and communication in the city. This comes under the form of an ever-changing network and of the description of the flows along its nodes and links. DUA takes this and reflects how it is possible to intervene in these flows, to create action, engagement, and transformation: Just like traditional acupuncture, it finds the "pressure points" on which it is possible to act to stimulate, inhibit, and suggest growth or the manifestation of limits, to create discussion and engagement, and to break or enhance existing loops and recurring patterns. The chapter will start off from the definition of Urban Acupuncture and from all its supporting theories and methodologies and then to the definition of DUA: its goals, methodological approaches, assumptions, and operative strategies. This will allow for attempting a dynamic—and polyphonic—definition of the well-being of the city and to describe the strategies according to which "pressure points" can be activated in creative ways, to provoke results, affecting wide areas of the population. A final section of this chapter explores the issues connected to the multiple forms of divide which can be present in this scenario and possible ways in which it is possible to confront with them: digital divides, cultural divides, age divides, and more.

At the end of the chapter, an interview with Marco Casagrande will highlight the connections between Urban Acupuncture and Digital Urban Acupuncture, and the roles of information and communication technologies in this.

Chapter 10 will present another case study, based on the analysis of Rome's cultural ecosystem which took place in 2013 with the support of the city administration. The process was designed to stimulate public engagement directed to promote participatory decision-making and policy-shaping processes and to foster the emergence of peer-to-peer self-organization patterns, to create novel forms of economies in the city's cultural environment. The Relational Ecosystem of the city was observed for 6 months, after which an Open Data source was released and used through an open laboratory which was accessible to citizens, administrations, organizations, associations, companies, and multiple types of cultural operators, which could learn how to use the Relational Ecosystem to achieve their goals. The chapter describes this experience, its results, and the directions in which it is currently heading.

Chapter 11 will expand the notions and concepts presented in the previous chapters by formulating a set of hypotheses about the implications of non-human engagement in these processes: What are the social, political, economic, and operative effects of including non-human agents in the Relational Ecosystem of the city, such as plants, trees, infrastructures, organizations, sensors, and devices? The chapter also points out the possibility to include different layers within the ecosystems, like in the case of organizations—for which their members appear in the ecosystem as both individuals and as the result of another ecosystem, the one of the organization—and in the case of distributed sensing networks—for example the natural environment of the city, which can be observed through disseminated sensors placed across the natural spaces of the city.

At the end of the chapter, an interview with Pier Luigi Capucci will highlight some of the more philosophical, critical and theoretical aspects which emerge when Relational Ecosystems form involving human, non-human and posthuman subjects, all communicating, relating and influencing each other through complex dynamics, for example through algorithms, organizational systems and control/power architectures.

Chapter 12 will present a case study coming from the city of Sao Paulo, in which a Real-Time Museum of the city was created in collaboration with SESC (the Brazilian federal agency in charge of managing the nation's museums and cultural installments) and Universidade Metodista. The museum was an iconic, engaging, inclusive experience of the Relational Ecosystem of the city. It has been designed to create awareness about the many layers of the Relational Ecosystem of the city and to provide a cultural shock to the population, to promote a radical transformation of the sense of possibility about the perception of what a metropolis can be and about the ways in which the concept of citizenship can be transformed, with citizens becoming positive, active, aware, and participatory agents of the city's civic life. The museum hosts a laboratory in which all members of society can learn how to use the Relational Ecosystem of the city, using DUA techniques, to pursue their goals and to create patterns of self-organization and intervention on the urban environment, supported by the city, regional, and federal administrations. The chapter narrates the methodologies, design processes, objectives, scopes, and expected impacts of the project, documenting the patterns, expectations, and future scenarios for this initiative, starting from a citizen-initiated research on Sao Paulo's water crisis.

Chapter 13 will present possible political and economic models which take into account the opportunities opened up through DUA and, in general, through the availability of wide areas of active citizens who are aware of the Relational

Ecosystem of the city and who can use it together with the Third Infoscape, to organize themselves, to create participatory decision-making processes, and to enact peer-to-peer dynamics in the city. These can best be synthesized as a shift in focus: away from the idea of the necessity of consensus and toward the idea of the possibility and feasibility for coexistence, valuing differences, and turning them into the creative energy driving the well-being of the city, establishing a parallel with the biodiversity of natural environments.

Chapter 14 will present some conclusions, highlighting the possible directions for further research.

Chapter 2
How to Use This Book

Abstract This chapter introduces a series of features which are characteristics of this book, which comes under the form of the *expanded book*. A series of special contents are disseminated throughout it, such as databases, videos, artworks, and interactive experiences, under the form of Internet resources which can be accessed through a variety of devices and which greatly enhance the experience of the publication. While this book is completely usable and accessible in its paper form, these characteristics allow to expand its experience.

This book assumes the form of what is commonly referred to as an "expanded book."

While being fully usable in its paper form, this book builds upon it, providing the possibility to access images, data sets, interactive experiences, digital interactive maps, interactive information visualizations, videos, and other materials. This allows for a deeper immersion in the subject and to have the availability of extended tools and information which can be used for the education purposes, further research, presentation, or even just to enjoy the wonderful images, interactions, and designs.

Here, we will describe them and organize them according to their type.

It may be useful to note how, regardless of the media, a form of each content has been produced for the paper version of this book, in order to make it fully usable. Even if you do not want to use your computer, smartphone, or tablet, you will be able to get the most out of this book through images, information visualizations, meaningful screenshots, and video captures.

Of course, some digital assets (such as datasets) have no direct paper counterpart: You will need to access the Internet addresses to download them.

Some of the datasets and visualizations will automatically update themselves over time. Kansas City's innovation ecosystem (Chap. 7) is one of them. When this happens, it will be explicitly mentioned, meaning that the specific information visualization and the database feeding it are updated over time: If you open it now, you will see the current data visualized; if you open it next year, you will see next year's data, as well as the current one. This can be an interesting opportunity to

S. Iaconesi and O. Persico, *Digital Urban Acupuncture*
DOI 10.1007/978-3-319-43403-2_2

show the evolution of data over time and may be useful for research, students' projects, or just to browse the developments of the described phenomenon.

In all of the following cases, each Internet link will be expressed using its textual representation (such as "http://www.human-ecosystems.com").

We will now detail the various types of expanded content which are included in this book.

2.1 Datasets

Datasets will show that a database is available for downloading at the specified Internet address. A short description of the data will be provided. The data will be relevant to the section in which it appears (most of the time, it will be the dataset which was used to obtain the results presented in the relative section of this book).

Data are provided in standard SQL format, which is suitable for importing from popular databases, such as MySQL or Postgres. Data are always commented directly in the files to give insights about the formats used within it.

With minor adaptations, all data can be exported to other formats, such as Microsoft Excel tables, CSV (comma-separated values), and more, for example, by first importing it into a database and, then, using the database's functions to export it to the desired format.

The datasets are downloadable under the form of ZIP file archives. Each archive contains two folders named "Scheme" and "Data": The former contains the definitions of the table structures of the database and the latter contains the actual data.

2.2 Videos

Multiple videos are included in this book. The links to these videos will be placed next to the relative video frames disseminated throughout this book. These videos can be accessed by typing the provided Internet address in a Web browser.

All of the videos use HTML5 players and MP4 formats, and thus, they should be accessible from any modern device.

2.3 Interactive Information Visualizations

Interactive information visualizations will allow to reach out into the images of the infographics shown in this book and actually use them to explore them using filters, different perspectives, specific queries, and points of view.

Their links will be placed next to the printed infographics and can be accessed by typing them in a Web browser.

All of the interactive experiences use standard HTML5 and are designed to be accessible from any modern device. Some of them require quite a lot of processing power, so they may be unsuitable for those devices which are able to provide little of it, such as older smartphones and tablet computers.

2.4 Interactive Maps

Interactive maps will allow you to zoom in/out, pan, and explore the geographical information depicted throughout this book.

Their links will be placed next to the printed infographics and can be accessed by typing them in a Web browser.

All of the interactive experiences use standard HTML5 and are designed to be accessible from any modern device.

2.5 Images, Artworks, and Other Works of Creativity

DUA has been extensively used by designers, artists, and other creative professionals to create engaging, rich interactive experiences, artworks, services, and beautiful visual representations and even to create generative symphonies, urban toys, and many more creative things. You will be able to access many of them—or of their representations—directly from this book.

Their links will be placed next to their printed representations and can be accessed by typing them in a Web browser.

All of the interactive experiences use standard HTML5 and are designed to be accessible from any modern device.

Chapter 3
Third X

Abstract Starting from De Certeau's Practice of Daily Life, the chapter describes how the theory of (Digital) Urban Acupuncture is connected with the concept of Third Space (Soja), Third Landscape (Clément), Third-Generation City (Casagrande), and Third Paradise (Pistoletto), arriving to a definition of Third Infoscape. The chapter ends with an interview with Gilles Clément.

When we experience territories, we create stories. We model these stories using mental maps. A mental map refers to one person's point of view perception of their own world and is influenced by that person's culture, background, mood and emotional state, instantaneous goals and objectives.

If we move along the streets of a city in a rush, trying to find a certain type of shop or building, our experience will be different than the one we would have had if we were searching for something else, or lazily strolling by.

Focus will change. We will see certain things and not notice other ones which we would have noticed otherwise. Some things we will acknowledge because they are familiar, common, or because we associate them to memories and narratives. Some will stand out because they react with some element of our culture or background. This process continuously goes on as our feelings, emotions, objectives, and daily activities change, creating the tactics according to which we traverse places and spaces, to do the things we do.

In the density of cities, this process happens for potentially millions of people at the same time. In his "the Image of the City," Lynch (1960) described cities as complex time-based media, as symphonies produced by millions of people at the same time in their polyphonic way of acting, moving, interpreting, perceiving, and transforming the ambient around themselves: a massive, emergent, real-time, dissonant, and randomly harmonic work of time-based art with millions of authors that change all the time.

In this, our mental maps—the personal representations of the city which we build in our minds to navigate them to fulfill our needs and desires—live a complex life as our perception joins into the great performance of the city.

S. Iaconesi and O. Persico, *Digital Urban Acupuncture*
DOI 10.1007/978-3-319-43403-2_3

Dissonance is the essence of the city itself and represents its complexity, density, and opportunities for interaction. Harmony represents the affordances, the things which are recognized and shared by different cultures. Those elements of the perceptive landscape onto which we can agree upon, which we recognize and attribute compatible meanings, allowing us to collaborate, meet, do things together. For example, Haken and Portugali (2003) have suggested a broad definition of landmarks to refer to any distinguished city elements that shape our mental map, or as Appleyard (1969), Golledge and Spector (1978) who have conducted studies about the imageability of urban elements not because of their visual stimulus but because they possess some personal, historical, or cultural meaning.

These features found within our mental maps enable the possibility to design the affordances of places and spaces. We can use the understanding of what is consistently recognized and understood to design the elements of space/time which will be able to describe to people what is allowed or prohibited, suggested or advised against, possible or imaginable. Lynch's concepts of legibility and imageability are closely related to Gibson's (1979) notion of affordances developed in his direct perception theory, according to which the objects of the environment can afford different activities to various individuals and contexts. And, again, in Haken and Portugali, all elements of a city afford remembering, as they shape in the mental maps in human minds.

These considerations may open up a series of remarkable observations and opportunities.

For example, in our cities the majority of these polyphonic understandings of place, space, time, and emotion are not readable. Or, when they are, they become readable for short periods of time, or in limited contexts, or for a limited number of people, or, maybe, only in case of exceptional occasions, such as riots or violent acts.

Nonetheless, gaining accessibility to these understandings could prove to be valuable for citizens, planners, designers, architects, entrepreneurs, public administrators as well as for artists and other creative professionals, as it would enable them to have the availability of a shared, inclusive source of knowledge, describing the life of the city with exceptional complexity, detail, and granularity.

In a further step—in the direction of citizen activation—we could imagine to make these types of understandings of the city widely known and also usable/performable, to enable people to express themselves (and their mental maps of how they perceive the world), and to act more effectively and powerfully.

Going in this direction, we could imagine to start our investigation focusing on three major axes: time, space, and relation.

3.1 History

Regarding time, we could start from the notion of history.

When we study history, we tend to imagine studying "large history": the history of the great people and great events/trends which transformed the course of time, wars, and societal transformations. Changes happened on a large scale, with treaties, alliances, and agreements which shaped the lives of entire populations.

This is, of course, not the only way in which we can research and study history. The term "history" itself is an umbrella term enclosing a rich variety of different approaches.

Among them is the possibility to study microhistory, the intensive historical investigation of a well-defined smaller unit of research (most often a single event, the community of a village, and a family or a person).

This might seem somewhat less relevant than history, as it could resemble an effort to focus on things which are of smaller importance, almost case studies that would, then, need to be framed into a wider context to be significant.

But if we think about it, this might well not be the case.

History is, of course, the result of the progression of large-scale transformations to the structures of human societies, their relationships, their disputes, and agreements.

But these large changes do not happen in a vacuum. They happen within human societies, which are made of human beings, and by their relationships, cultures, imaginations, desires, and expectations.

So it is possible and valuable to view the study of history also as the possibility to "ask large questions in small places or contexts," as hypothesized by Joyner (1999).

To try to explore the conditions in which these large-scale events and transformations actually took place, through people and the mutation of their daily lives, of their cultures and desires.

And that is precisely what happened when historians started to understand that certain "political events and social realities" could not be explained adequately by existing macrohistorical models, as highlighted, for example, by Levi (1991).

In essence, historical histories did not account for the experiences of all members of the event, society, or culture being studied. As a result, microhistorians have made a point of viewing people not as a group, but rather as "individuals who must not be lost either within the historical processes or in anonymous crowds" (Iggers 1997).

Microhistorians have attempted to formulate a history of everyday life (Brewer 2010).

3.2 Everyday Life

In his "The Practice of Everyday Life" De Certeau (1984) transformed the study of "everyday life," shifting it away from the study of popular cultures and from the research about the social and political struggles which happen with the daily forms of resistance to the regimes of power, in an attempt to outline the way individuals unconsciously navigate everything, from city streets to literary texts.

This approach leads to an interesting distinction among the strategies and the tactics.

The idea of strategies is linked to the one of institutions and to the structures of power, describing and producing the prescriptions (the codes) according to which the elements of reality should be interpreted. They are the official rules of society: the laws and regulations, the official usages of objects and spaces of the city. They are enacted by encoding, by putting objects and places on maps with precise legends (or codes), or by establishing boundaries and borders.

On the other side are the tactics, referring to people and the ways in which they continuously surf the strategies in unexpected ways, navigating them according to their cultures, desires, urgencies, and imaginations. People constantly perform the environment producing their own interpretations of reality, using objects and moving through cities in ways that are tactical and never fully determined by the plans of organizing bodies.

People fundamentally and continuously break the codes established by the strategies, enacting their tactics and, thus, reprogramming the environment, and adding new codes onto it, established by acts of "making" and of "performing," by unpredictably changing their trajectories while moving through urban space, by changing the way in which they use a certain object, and by mutating the way in which a certain space is lived.

In the chapter "Walking in the City," De Certeau writes:

The ordinary practitioners of the city live "down below", below the thresholds at which visibility begins. They walk—an elementary form of this experience of the city; They are walkers, Wandersmänner, whose bodies follow the thicks and thins of an urban "text" they write without being able to read it. These practitioners make use of spaces that cannot be seen; their knowledge of them is as blind as that of lovers in each other's arms. The paths that correspond in this interweaving, unrecognized poems in which each body is an element signed by many others, elude legibility. It is as though the practices organizing a bustling city were characterized by their blindness. The networks of these moving, intersecting writings compose a manifold story that has neither author nor spectator, shaped out of fragments of trajectories and alterations of spaces: in relation to representations, it remains daily and indefinitely other.

And, later:

Their story begins on ground level, with footsteps. They are myriad, but do not compose a series. They cannot be counted because each unit has a qualitative character: a style of tactile apprehension and kinesthetic appropriation. Their swarming mass is an innumerable collection of singularities. Their intertwined paths give their shape to spaces. They weave places together. In that respect, pedestrian movements form one of those "real systems whose existence in fact makes up the city." They are not localized; it is rather that they spatialize.

According to this idea, strategies and tactics each produce distinguishable parts of the city. The first is top-down and is relatively static, relating to the institutionalized, bureaucratic, legal, and administrative codes which describe the spaces of the city. The second is bottom-up, emergent, dissonant, in real-time, describing the desire and visions of the city-practitioners (the performers), written on the cities through their bodies and their actions within the city.

The first represents a top-down form of information and knowledge. The second is bottom-up.

The first is mainly static, and highly readable through the apparatus of signage, visual encoding, and images produced by administrations.

The second one is dynamic, ever changing, multiple, and polyphonic; is below the threshold of readability as it is drawn through the bodies of city dwellers; and is ephemeral, lasting only a few instants.

According to De Certeau, this form unpredictable creativity describes a space, in which revolutionary potentials exist, in which individuals individualize culture, and turns elements of the popular in their own, reappropriating them.

This new space comes together in the concept of Third Space.

3.3 Third Space

The Third Space breaks the dialectic between the First Space (home) and the Second Space (the workplace).

In "The Location of Culture," Bhabha (1994) described hybridity and the Third Space from a postcolonial perspective. In the book, he extended Walter Benjamin's work by describing Third Space as differential temporal movements within the process of dialectical thinking and the supplementary or interstitial "conditionality" that opens up alongside the transcendent tendency of dialectical contradiction. According to him, the Third Space is a hybrid, interstitial, emergent space which is constructed through a process of differentiation and multiplication.

In 1996, political geographer and urban planner Edward Soja published his "Thirdspace," drawing inspiration from postmodern analysis influenced by Henri Lefebvre, Michel Foucault, and Bhabha. In his view, Third Space "breaks the Firstspace-Secondspace dualism and comprises such related concepts as 'place, location, locality, landscape, environment, home, city, region, territory and geography' … that attempts to come to terms with the representational strategies of real and imagined places." Soja proposes a "trialectics of spatiality" which is a process, a dynamic force, and "recombinational and radically open."

According to Soja in the Third Space:

"*everything comes together… subjectivity and objectivity, the abstract and the concrete, the real and the imagined, the knowable and the unimaginable, the repetitive and the differential, structure and agency, mind and body, consciousness and the unconscious, the disciplined and the transdisciplinary, everyday life and unending history.*"

Third Space is a radically inclusive concept, in which the strategies exist together with the tactics which, thus, gain visibility and perceivability, enabling the contestation and renegotiation of boundaries and cultural identities.

This is a process which is very similar to Homi K. Bhabha's theory of cultural hybridization, in which "all forms of culture are continually in a process of hybridity" that "displaces the histories that constitute it, and sets up new structures of authority, new political initiatives… The process of cultural hybridity gives rise to something different, something new and unrecognizable, a new area of negotiation of meaning and representation" (Rutherford 1998).

Over time and situations, the concept of Third Space has been used in different contexts to describe the possibility of establishing relational, emergent dynamics in order to create multimodal, tactical processes which would allow to overcome the organized architectures of power and to conceive novel dialogues.

Whether it is in the whispering corners or taverns in which the oppressed plot their liberation, or the liminal spaces in which they are able to come together with their oppressors, embodied in their particularity; whether in education, where students' cultural capital merges with the official curricula (Maniotes 2005), and where multiliteracies take form (New London Group 1996; Skerrett 2010); whether in the domestic sphere, as an additional space beyond family and the spaces for civic expression, the places in which people let their "real" selves show (Walsh 2006; Whitchurch 2008; Hulme 2009); or in sports associations (Ruddock 2005), bars, and nightclubs (Law 2004).

Thus, the Third Space is a space for open opportunity through the recognition (to see) of "possibility" in the "other," and in the tactics that are expressed in space. Opportunity exists if our gaze can become educated to see the tactics and to learn to negotiate their meaning.

3.4 Third Landscape

The Third Space is a place of open, shared possibility and, to be recognized as such, it needs for an "education of the gaze," a new aesthetic sensibility.

Its emergence reminds of Gilles Clément's Third Landscape.

When Gilles Clément described the Third Landscape (2005), he described it as:

- *an undetermined fragment of the Planetary Garden—designates the sum of the space left over by man to landscape evolution—to nature alone. Included in this category are left behind (délaissé) urban or rural sites, transitional spaces, neglected land (friches), swamps, moors, peat bogs, but also roadsides, shores, railroad embankments*, etc. *To these unattended areas can be added space set aside, reserves in themselves: inaccessible places, mountain summits, non-cultivable areas, deserts; institutional reserves: national parks, regional parks, nature reserves.*

Compared to the territories submitted to the control and exploitation by man, the Third Landscape forms a privileged area of receptivity to biological diversity. Cities, farms and forestry holdings, sites devoted to industry, tourism, human activity, areas of control, and decision permit diversity and, at times, totally exclude it. The variety of species in a field, cultivated land, or managed forest is low in comparison with that of a neighboring «unattended» space.

The Third Landscape is the part of the natural environment that grows in-between bricks and stones; it is the grass that lives between train tracks; it is the natural space that finds its life in the cracks of the walls, or in the places of our cities to which we do not pay much attention.

It is the natural space of our cities which has not yet been encoded. It is not found in the flowerbeds and hedges which our city administrations define through borders and limits: please keep off the grass, this is a bureaucratically instituted flowerbed.

From an ecological point of view, the larger part of the biodiversity in our cities is found in the Third Landscape.

From this point of view, the Third Landscape can be considered as the genetic reservoir of the planet, the space of the future.

Gilles Clément's "Planetary Garden" is one of the most suggestive answers to the mutation of the definition of urban space. Planetary Garden is to economic and urban globalization what urban gardens were to the cities of the nineteenth century: the latter represented the closed or tightly schemed design of urban architecture and layout, while the former represents the connective, fluid, mutating texture of the globally interconnected city. The Planetary Garden is the garden of the global city.

The Third Landscape is a connective fabric composed of residual spaces that tend to take a liquid state, never preserving shape, resisting governance. Classical preservation or environmental conservation tools such as surveillance, protection, and the creation of limits and borders cannot apply to the Third Landscape without destroying its characteristics. As Clément (1999) writes, the Third Landscape is "not property, but space for the future." An idea of space goes beyond the ideas of landscape as a place for identity, being used as an asset for local societies, and as a strategic tool for memory.

An idea of space exemplifies the possibilities of the contemporary world: a multiplication of narratives; the holistic perception of ecosystems; the possibilities and richness offered by disseminated, interstitial, emergent, mutating, temporary, polyphonic environments; the end of dualistic approaches.

As Barrell (1980) spoke about "the dark side of the landscape" while pointing out the imposition of a point of view of a single social class, with Clément we could speak about a "light side," for the Third Landscape is not an exclusive model but an inclusive one: "*a shared fragment of a collective consciousness.*"

It is based on a planetary remix (*brassage*) which is at the origin of the current richness of ecosystems. These dynamically mutating spaces embody the presence of multiple agencies forming the city from points of view that are architectural, political, economic, poetic, activist, industrial (Di Campli 2005): new forms of nature that emerge by instantaneously creating interstitial ecosystems that flow with the story of the city, describing a real-time, syncretic map that develops together with the creation of new areas for residences, industry, commerce, business, culture and entertainment, and with the death, abandonment, and decay of the previous ones, as a geography of the mutation of the city (Iaconesi 2012, Leaf++).

Clément talks about the necessity of training our gaze into recognizing and understanding the Third Landscape. This requires a new possibility for vision and knowledge dissemination in urban natural environments, a renewed sense of aesthetics, and a morphed sensibility for the possibilities for interaction and communication offered by our surroundings.

This is a potentially revolutionary point of view, as it alludes to the possibility to achieve the perception of these emergences, and the possibility to transform them into a form of shared knowledge.

A similar discourse could be imagined for the Third Space: what if the emergent history of tactics could become a source of shared knowledge? What if the progressive sedimentation of this knowledge, its continuous formation of ever changing and evolving ruins, layer after layer, could become accessible and readable, through sensibility and *sense-ability*, and through a novel form of aesthetics to stimulate both perception (attention) and awareness, to describe the progressive history of daily life: a stratified, accessible, perceivable, usable, continuously evolving microhistory?

3.5 Third-Generation City and Urban Acupuncture

What is a ruin?

A ruin is the progressive reunification of objects and architectures to nature. As buildings grow older, the actions of natural agents, of human beings, and of the Third Landscape mutate them, bringing them into a different form: more organic, and systematically integrated into the natural environment.

In a way, nature and human beings ruin buildings, transforming them into ruins.

From a different point of view, the actions of human beings and nature bring buildings into a different state, transforming them into ruins, providing evidence of the history of humans' and nature's interventions on architectures, of the patterns according to which they have been used daily. From this point of view, ruins expose the history of the natural (and human, as integrated in nature) environment and of its daily life.

Ruins are, to all effect, a history and a source of knowledge and of information, enacted through the layering processes of the results of the actions of human beings and of natural agents.

From Marco Casagrande's definition (2013):

Third-Generation City is the organic ruin of the industrial city.

Third-Generation City is true when the city recognizes its local knowledge and allows itself to be part of nature.

And (Casagrande 2013a, b):

The Third-Generation City is the industrial city ruined by the people—human nature as part of nature.

Like a weed creeping into an air-conditioning machine the industrial city will be ruined by rumors and by stories. The common subconscious will surface to the street level and architecture will start constructing for the stories—for the urban narrative. This will be soft, organic and as an open source based media, the copyrights will be violated. The author will no longer be an architect or an urban planner, but somehow a bigger mind of people. In this sense the architects will be like design shamans merely interpreting what the bigger nature of the shared mind is transmitting.

This last definition is specifically interesting for all our discussion: the image of the layering of the subconscious, of the stories and narratives produced by people emerges as a novel (un)building material which is capable of preserving history and knowledge, by transforming spaces, whose authors will no longer be architects or planners, but people themselves.

The Third-Generation City is envisaged to be an organic layer that promotes alternative modes of living as well as narratives, or "urban rumors."

The Third-Generation City is a form of knowledge: just like rumors, it is formed by precious knowledge which is ordinarily nonreadable, not perceived.

As in the case of the Third Landscape, the gaze needs to become educated in order to recognize this kind of knowledge, this stratification, which becomes a new kind of aesthetics, a new form of perception for possibility and opportunity: an open space for the future.

It is imaginable to acknowledge the need for this education process and, thus, to imagine possibility to learn to see the city as a whole, as a body, which includes both architectures and their emergent, interconnected layering with the histories and knowledge of the daily lives of human beings and nature.

This body would not be static, with continuous, emergent flows of knowledge and information taking place throughout it.

This would enable, for example, the visions of architect Künnapu's (2007) theory of energy center architecture aiming in tuning the urban condition into a network of spiritual layers, as well as architect Marco Casagrande's theory or Urban Acupuncture in which the cities are treated punctually as energy organism toward an environmentally (and socially) sustainable development: the possibility to perceive "rumors," or, better, the myriads of syncretic expressions of the lives in the city, which would form a connective tissue.

This connective tissue would represent the mutating tensions, agreements, conflicts, oppositions, harmonies in the city, as they manifest themselves through flows of energy (intended as the capacity to perform work, as the driver of human action through the intertwined microdynamics which relate human beings, infrastructures, timetables, objects, relationships and more, driving us in the things we do every day).

We could use this concept (the idea of the flows) to establish a parallel with traditional Chinese Acupuncture, in which selected points and nodes are carefully picked along the meridians of the human body's energy flows to insert needles which would intervene on them. If we translate this concept into an urban context, we could imagine performing Urban Acupuncture, in which the knowledge and understanding of the flows of communication, information, and knowledge in the city are used to "insert needles," to perform localized interventions whose small scale is focused in ways which are able to stimulate these flows, in order to activate them and, thus, drive to action wide areas of the population (and their rumors, tensions, conflicts, agreements, desires, expectations) to transform larger urban contexts.

The city, in this case, would be viewed as multidimensional sensitive energy organism, a living environment. Urban Acupuncture aims to address this nature, stimulating its sensitivity to understand the energy flows of the collective chi beneath the visual city and reacting on the hot spots of this chi (Miller 2011).

Urban Acupuncture sees the city as a complex, ever mutating body, with narratives, emotions, information, and knowledge exchanges as its main meridians for energy flows, also bearing some similarities to the new urbanist concept of Tactical Urbanism (Garcia and Lyndon 2015). The idea focuses on local resources rather than capital-intensive municipal programs and promotes the idea of citizens installing and caring for interventions, and on the opportunities which come with the ability to understand the drivers which can stimulate citizens to activate themselves in these caring processes, in a state of participatory performance whose drivers are aesthetic and desiring. These small changes, proponents claim, will boost community morale and catalyze revitalization (West 2011).

3.6 The Info-Body of the City: The Third Infoscape

As we have seen so far, the idea of microhistory allows us to focus onto the personal stories of people, describing territories not only in terms of the large-scale events and trends which happen in (or to) them, but allowing for a multitude of points of view emerge, the histories of people's daily lives, which can be observed to make sense of the larger phenomena.

These stories form the tactics, described by De Certeau, which, together with the strategies, encompass the dialectic confrontation between the top-down and bottom-up encodings of cities. The first ones are static and prescriptive, establishing strict codes and boundaries. The second ones are dynamic and emergent and describe the performative practices of city dwellers, in their reinterpretation and reappropriation of the spaces of the city. This process happens in the Third Space, as described by Soja.

In a parallel with Clément's Third Landscape, we have seen the ways in which the Third Space can be used as the space for emergent opportunity in the city: an inclusive, possibilistic, and accessible open space in which it is possible to define new, emergent codes, at multiple levels and according to different directions. To do this, new forms of aesthetics and perceptions must be achieved, to be able to perceive the Third Space/Third Landscape, to see and interpret it as the open space for opportunity and for a possibilistic description of the future.

With the Third-Generation City, we have seen how to integrate all these levels using the idea of the ruins, in which tactics stratify on top of strategies, transforming them. This layering represents the effects of nature and of human daily lives on the spaces described by the strategies, their histories, and narratives.

This, in turn, describes the city as a body, in perpetual dynamic evolution, in which this emergent process describes the flows of expression, emotions, information, and knowledge: the energies of the city.

These flows, which we can imagine as similar to the ones described in traditional acupuncture, are the source of knowledge and information we can use to apply Urban Acupuncture, acting on the nodes of their meridians to liberate and enhance them and, thus, producing larger effects through small interventions.

All of this process we have just described relies on the energies of the city which are represented by expression, emotion, information and knowledge exchanges, and on their possibility to flow freely, and to leave evidence of their (micro)history to be transformed into accessible forms of awareness, wisdom, insights, enlightenment, and performance.

In current times, much of these energies assume digital forms.

We have learned to use mobile devices, ubiquitous technologies, social networks, and other ubiquitous forms of communication to work, collaborate, make decisions, express our feelings, learn, communicate, establish relationships, and consume (Iaconesi and Persico 2012a, b; Urbanverse 2011).

For this, we can imagine to extend the reach of previous definitions from the context of physicality to the digital one, to attempt a starting definition for the (First, Second, and) Third Infoscape. The First Infoscape would refer to the information and knowledge generated within nature; the Second Infoscape would refer to the information and knowledge generated in the industrial city (the Second-Generation City, the city of infrastructures, of transactions, of sensors, of bureaucracy, and of administration); and the Third Infoscape would refer to the information and knowledge generated and exchanged through microhistory, through the progressive, emergent, and polyphonic sedimentation onto the city of the expressions of the daily lives of city-practitioners, through their spontaneous, unconscious, or conceptual performance of the city.

By creating this definition and focusing on it, it would be, then, possible to focus our attention onto the Third Infoscape—together with the First and the Second—to create a novel kind of sensibility, perception, and awareness. And with this new form of sensibility, it would be imaginable to form new modalities for observing and understanding our cities through citizens' expressions happening in digital domains, in inclusive, spontaneous, performative ways. New kinds of Urban Acupuncture interventions could be thought in this way, based on the energy flows of the city, expressed through the digital domains which are now a fundamental part of our daily experience, inseparable from the physical one, and mutually influencing each other.

To achieve this, we would need to form a new aesthetic (referring to the concept of perception) sensibility, to see the Third Infoscape, and to recognize it as an inclusive space for opportunity, in the same sense pointed out by Clément when dealing with the Third Landscape.

3.7 Third Paradise

In this sense—in the sense of the need for the emergence of a new aesthetic to become truly able to benefit from the "Third X," from all of these Thirds—another concept comes of use, which is Michelangelo Pistoletto's concept of the Third Paradise (2010).

In his definition of the Third Paradise, Pistoletto describes:

The project of the Third Paradise consists in conducting artifice, science, technology, art, culture and politics to give back life to Earth. Third Paradise indicates the passage to a new level of planetary civilization, which is essential to ensure the survival of humankind. The Third Paradise is the new myth which brings everyone to assuming a personal responsibility in this epochal scenario. With the New Infinite Sign we draw three circles: the middle one if the generative womb of the Third Paradise.

The Third Paradise, as demonstrated through its symbol, an infinite sign with a third loop, is a Third Space. It is a space in the middle, in-between, interstitial, in which the idea of the conflictual and destructive relationship between nature and technology vanishes, enabling both to become part of society. As Pistoletto says, the driver for this mutation is people's sense of personal responsibility and relationality, expressed through this new, dedicated aesthetic. The Third Paradise is a state of dynamic, participatory performance inspiring a sense of responsibility and collaboration, as well as a participatory form of knowledge.

In this sense, the Third Paradise gives us an idea for what the new aesthetics and sensibilities for the process discussed in this book could look and feel like.

3.8 An Interview with Gilles Clément

Gilles Clément *is a French gardener, garden designer, botanist, entomologist and writer. He has gained attention for his design of public parks in France, such as Parc André-Citroën. In 1998, he was the recipient of France's National Landscape Prize. Since 1977 he has developed his "moving garden" (le jardin en mouvement) at La Vallée, Creuse. He coined the terms "Third Landscape" and "Planetary Garden."*

What does a "gardener for a garden without a form" look like? What does he or she do?

In a garden without a form the gardener focuses his interest in diversity and in the vital exchanges among the beings which compose this diversity. We can imagine giving a form to the garden, but it will not be necessary to ensure a gardening process which privileges life.

The garden is an enclosure which is destined to protecting the «best». But the idea of best changes over time. The message inscribed in the garden changes with it. While we need a precise form to valorize water in the hispanic-mauresque gardens, power thanks to perspectives in classical gardens, point of view—belvedere which face a dramatized nature in the romantic gardens, today's garden doesn't need a design to confront with the question of life, which is today's fundamental issue.

We have asked you the previous question because in the informational landscape of cities we find ourselves in a similar situation. When it comes down to information and knowledge, the city is a "garden without a form": it evolves, spontaneously, with flows and apparent chaos which are the result of complex relational dynamics. The beauty of this garden is in its richness and in its energy, and in the fact that it constitutes a map of the city which goes beyond buildings and roads, and arrives at describing the syncretic city, made of relations, exchanges, communication, emotions.

In your practice, how important are maps? How important are maps for a "Planetary Gardner"?

In our practice maps and plans play the role of framing the spaces which are defined and limited which correspond to the territory which is the object of research and intervention. But the landscape does not have anything to do with the territory of intervention. The gaze does not stop at the cadastral boundaries. Landscape cartography is always meant as more extended than a territorial map. Its design corresponds to a series of boundaries which vibrate, and in which the straight line is excluded.

The maps of the Third Landscape and their evolution in time and context may be interpreted as maps of the biodiversity in the city. By making the parallel, we say that the maps of the Third Infoscape represent the cultural diversity in the city, as it flows and evolves in space, time and context.

How can these representations, in your opinion and practice, be used to foster awareness about the importance of diversity, and as a tool (for citizens and professionals) to understand the value of diversity, to be used to act in the city and promote the well-being of its citizens and of the environment?

In the city, the cartography of the spaces of the Third Landscape can be read in relationship with routes of circulation and buildings, in a geometry defined by urbanism. But the interior of these forgotten spaces doesn't have a form. It is hard to valorize these spaces using policies. The only possibility is to explain the fundamental role of the biodiversity which is welcomed in these ecosystems: it is thanks to this diversity found in these spaces that it is possible to «garden» the spaces for human intervention in order to act on both the economic and ecologic planes.

You often speak about the need for new aesthetics, to understand the value and importance of the Third Landscape. The "abandoned spaces" of the city

are often considered in a negative sense, not as a "space for possibility," or an "open space for the future." This also happens for data and information in the city, where the dimension of "gossip," "chitchat" and "conversation," which constitutes a large part of the Third Infoscape, is often seen as unimportant, irrelevant, meaningless. And administrators, planners, architects, designers and even citizens refer to more encoded forms of data, information and knowledge to make informed decisions about the present and future of the city. Instead, through "gossip," "chitchat" and "conversation" we can discover many things about the city and its inhabitants: their desires, wishes, expectations, cultures, visions, emotions and imagination.

How do you create new aesthetics for the Third Landscape? What is the role of artists, poets, performers and narrators in the city you imagine?

A new aesthetic of landscape which starts from the Third Landscape presupposes an expansion of the freedom for biological expression. To accept one must comprehend. To comprehend access to knowledge of the living is necessary. Bad herbs, weeds, don't exist. Herbs exist, each with a name, each with a role in the ecosystem. Each bound to an insect, or a bird, or to another plant, and these connections are often intertwined. Once we have understood the importance of these interconnections in the life of a being we comprehend how it uses this diversity of connections to live, we accept diversity, the expansion itself seems «inaudible».

In your work you often mention the need for transformation of the rules and the intentions about the ways in which we consider intellectual property and property in general. The Third Landscape is "open" and "accessible" and "inclusive." This is true also for data, information and knowledge. It seems to us that the current Open Data—and related—actions happening in cities of the world go in this direction. But, at the same time, they lack the focus on the high quality relational environment which is needed in order for citizens to be able to create meaning and shared objectives.

What kind of transformation of intellectual property and property in general do you envision? How does it manage to create a high quality relational environment?

Property is, historically, enclosing a «commons». Nature is composed of a systemic whole of commons. We must have free access, without patenting in favor of laboratories and without interdiction for the citizens. Today, to ensure access to the commons, we must surround ourselves with lawyers and jurists to prevent multinationals from pressuring legislators to create laws which are unjust and which intend to interdict access to the commons.

You often speak, in your work, of "controlled form" as a thing we needed in the past to fight the fear of the unknown and, on the other hand, the ways in which the friches are the place of life, the places in which invention arises. Could this lack of form be really the "blank slate" which we need to start judging things "not based on their form, but for the joy of existence they express"?

Form has always its visual, aesthetic importance, which is also functional pedagogically. For example: to valorize a space of formal confusion such as a fragment of the Third Landscape in which different herbs interweave with each other, it could be of great interest to create a design in which creates the perfect setting using mowed grass just to showcase the «tableau», the painting, to add value to it. Furthermore, this type of intervention means that «it is that which is inside the painting which we desire, it is not something forgotten by the gardener».

References

Appleyard D (1969) Why buildings are known. Environ Behav 1:131–156

Barrell J (1980) The dark side of the landscape: the rural poor in English painting, 1730–1840. Cambridge University Press, New York

Bhabha HK (1994) The location of culture. Routledge, London

Brewer J (2010) Microhistory and the histories of everyday life. CAS e-SERIES, 5

Casagrande M (2013a) Cross-over architecture. Available via Epifanio. http://www.epifanio.eu/nr9/eng/cross-over.html. Accessed 12 Nov 2015

Casagrande M (2013b) Third generation city. Available via Casagrande Lab. http://casagrandetext.blogspot.it/2013/10/third-generation-city.html. Accessed 12 Nov 2015

Clément G (1999) Le jardin planétaire. Reconcilier l'homme et la nature. Albin Michel, Paris

Clément G (2005) Manifesto del terzo paesaggio. Quodlibet, Macerata

De Certeau M (1984) The practice of everyday life. University of California Press, Berkeley

Di Campli A (2005) Review of the "Manifesto del terzo paesaggio." Available on Architettura.it. http://architettura.it/books/2006/200602004/. Accessed 12 Nov 2015

Garcia A, Lyndon M (2015) A tactical urbanism how to: short-term action for long-term change. Island Press, New York

Gibson JJ (1979) The ecological approach to visual perception. Houghton-Mifflin, Chicago

Golledge RJ, Spector A (1978) Comprehending the urban environment: theory and practice. Geogr Anal 10:403–426

Haken H, Portugali J (2003) The face of the city is its information. J Environ Psychol 23(4):385–408

Hulme R, Owens A, Cracknell D (2009) Learning in third spaces: developing trans-professional understanding through practitioner enquiry. Educ Action Res 17(4):537–550

Iaconesi S (2012) Leaf++. Available via Art is Open Source. http://leaf.artisopensource.net/. Accessed 12 Nov 2015

Iaconesi S, Persico O (2012a) ConnectiCity: real-time observation and interaction for cities using information harvested from social networks. Int J Art Cult Des Technol (IJACDT) 2(2):14–29

Iaconesi S, Persico O (2012b) The co-creation of the city. In: ECLAP 2012 conference on information technologies for performing arts, media access and entertainment 62

Iggers G (1997) From macro-to microhistory: the history of everyday life. In: Historiography of the 20th century. University Press, New England

Joyner CW (1999) Shared traditions: southern history and folk culture. University of Illinois, Urbana

Künnapu V (2007) Temples and towers. Available via Vilen Künnapu. http://www.vilenkunnapu.pri.ee/templid_sisu_eng/temples.html. Accessed 12 Nov 2015

Law L, Amin A, Thrift N (eds) (2004) Negotiating the bar: sex, money and the uneasy politics of third space. The Blackwell cultural economy reader. Oxford Blackwell, Oxford

Levi G, Burke P (eds) (1991) On microhistory. In: New perspectives on historical writing. Pennsylvania State Press, University Park

Lynch K (1960) The image of the city. MIT Press, Cambridge
Maniotes LK (2005) The transformative power of literary third space. Doctoral dissertation. School of Education, University of Colorado, Boulder
Miller K (2011) Urban acupuncture: revivifying our cities through targeted renewal. MSIS 9/2011
New London Group (1996) A pedagogy of multiliteracies: designing social futures. Harvard Educ Rev 66:60–92
Pistoletto M (2010) Il terzo paradiso. Marsilio, Venice
Ruddock A (2005) Let's kick racism out of football—and the lefties too!: responses to Lee Bowyer on a West Ham. J Sport Soc Issues 29(4):373
Rutherford J (1998) The third space. Interview with Homi Bhabha. In: Identity: community, culture, difference. Lawrence & Wishart, London
Skerrett A (2010) Lolita, facebook, and the third space of literacy teacher education. Educ Stud 46 (1):67–84
Soja EW (1996) Thirdspace. Blackwell, Malden
Urbanverse (2011) Urbanverse. Available via Urbanverse. http://urbanverse.net/21st-century-cities-c-is-for-co-creation/. Accessed 12 Nov 2015
Walsh K (2006) British expatriate belongings: mobile homes and transnational homing. Home Cult 3(2):125
West D (2011) 'Urban acupuncture' touted for cash-strapped cities. New Urban Network 7/2011
Whitchurch C (2008) Shifting identities and blurring boundaries: the emergence of third space professionals in UK higher education. High Educ Q 62(4):377

Chapter 4
Ecosystems in Cities

Abstract Starting from the definition of the Third Infoscape, this chapter introduces a vision of cities' ecosystems as a network of interconnected human and non-human entities establishing flows of communication, information, knowledge, emotion, and opinion. Cities are shown as a polyphony of expressions interweaving human and non-human communication, sensors, data, and information. The issue of legibility of tactics is also introduced in this chapter, as well as the concept of the necessity to gain better understandings about the myriads of microhistories of all the members of ecosystems to comprehend it. The chapter ends with an interview with Carlo Ratti.

4.1 Reading the City

There are many informational elements which we experience when we interact with cities.

Traditionally, they were physical: architectures, visuals, signage, sounds, and smells.

In his "Concise Townscape," Cullen (1971) provided a seminal description of the modalities which can be used to plan and design the shape of the city in order to stimulate emotional and interactive responses. The book emphasizes the role of personal experience in urban landscapes. This concept is remarked by highlighting the many ways in which we actually create places, under the form of mind maps: serial vision, juxtaposition, and immediateness. The analysis of the visual elements which allow us to comprehend public space and to understand its usages and rules brings us to the exploration in which city dwellers spatially perceive information in the continuous dialogue between the actual visual composition of the environment and the one which emerges as they traverse the city. In this kind of analysis, juxtaposition and immediateness (then declined as legibility or accessibility) of the elements of spatial information become emotional and operational actuators of the interactions between the city and its users (Cullen 1966, 1968).

S. Iaconesi and O. Persico, *Digital Urban Acupuncture*
DOI 10.1007/978-3-319-43403-2_4

Kevin Lynch, in "The Image of the City" (1960), started from a few case studies to explore the legibility of the city, formulating the hypothesis according to which people construct mental maps used to traverse, interact, and relate with urban spaces and the population. The possibility to collect and compare these mental maps enables to comprehend how visual, perceptive, cognitive, and psychological information in urban territories contributes to forming models of people's interactions and relations.

These and other texts—and the practices connected to them—tend to account for physical elements of reality and include the symbolic and cultural dimensions.

In "Learning from Las Vegas," Venturi (1977) moved beyond that. By observing the surreal and suggestive visions of the city of casinos in Nevada, he built an image of cities as expression of a society which progresses toward a reality constructed not only through symbols, but also through reproductions and representations.

This information comes under the form of iconographic instances and constitute a series of stratifications—of layers—on the territory: an immaterial geography which contributes to the modeling of citizens' mental maps and, thus, to the ways in which they act and react, just as the physical shapes, the sounds, smells, and tactilities.

Aldo Rossi, in conflict with Venturi's "Complexity and Contradiction in Architecture" (2002), merged a value and time-based analysis to the symbolic and postmodern perspective, alluding to the possibility of imagining the city as something which could be constructed with the flowing of time and the superimposition of the stories of its inhabitants, referring to the opportunity to build the place of the dynamic preservation of the collective memory (Rossi 1966).

These considerations may be added to the ones by Lefebvre, De Certeau, and Soja, among others, according to whom the construction of space and, thus, its communication and interaction are composed of a performative aspect which is polyphonic, and in which the microhistories of people (according to the definitions by Levi and Iggers) dynamically and generatively interweave in the continuous recomposition of urban space.

The Third Space is a radically inclusive concept, in which strategies and tactics coexist polyphonically and, thus, where the latter achieve visibility and legibility, allowing for the liberation, contestation, and renegotiation of boundaries and of cultural identities.

4.2 Ubiquitous Information

Data, information, and knowledge are ubiquitous.

They are in the shapes of buildings, in streets, and in urban furniture; in the forms of the paths chosen by city dwellers to traverse spaces and places; in signs, symbols, images, and icons; in colors; in the smells and sound we feel while we are in the city; in the skyline; in objects which are near, and in those we see at the horizon; in the memories which we associate to places, objects, and contexts; and in those memories which other people described to us, as we remember them, in precise ways, or not.

All of these elements—whether they are physical, immaterial, and ephemeral—and all of these sources of data, information, and knowledge contribute to give shapes to our actions, to our performance of the city.

Ubiquitous technologies and digital networks largely augment our possibility to generate data, information, and knowledge, to experience them, everywhere and at all times, and, thus, to shape our urban performance.

Smartphones, sensors, services of all kinds, interactive screens, urban screens, media facades, augmented realities, social networks, and natural interaction systems activated by gesture, voice, and movement and, even indirectly, through data, information, and knowledge which we come about through digital technologies: These and other devices, interfaces, and services allow us to transform each action in data and information, whether we realize it or not, consciously or unconsciously. It is the flipping on or off of a light switch; the passage within the field of view of a security camera; the querying for some train information using an interactive billboard; the purchase using a credit card; the usage of an app on our smartphone to find a restaurant we like; the notification on the health condition of one of our loved ones, generated through a biometric sensor; or the many other available possibilities through which any gesture action can be transformed into data and information.

The life of these data and information is complex.

Some of them are archived, according to various modalities and degrees of persistence, until they are needed.

Other are directly used, to execute services and processes.

For example, if *person A* desired to execute a purchase using a payment system on their mobile phone (maybe using Bitcoins[1]), a few clicks or touches on the device would be sufficient to execute the necessary data checks to authorize the transaction (generating more data).

For example, a *friend of A* could find himself in the shop and, realizing that he left home without cash or credit cards, could have asked *A* to execute the purchase for him, promising to give back as soon as possible. The *A* subject could be thousands of miles away from the store and, nonetheless, execute a physical purchase in that location.

Other modalities are possible, ranging from the more simple and legible ones (such as swiping an RFID[2] card in the immediate proximity of a reader to access the subway or to pass a tollbooth), to the more complex and opaque.

[1]Bitcoin is a digital currency created in 2009, which allows to execute anonymous economic transactions. For more information, it is possible to look at: http://it.wikipedia.org/wiki/Bitcoin.

[2]RFID means radio frequency identification and indicates a series of technologies which can mark objects with tags which can be read and updated automatically by dedicated readers in their immediate vicinity. The tags often come under the form of small stickers containing a passive radio frequency circuit. For more information, it is possible to look at: http://it.wikipedia.org/wiki/Radio-frequency_identification.

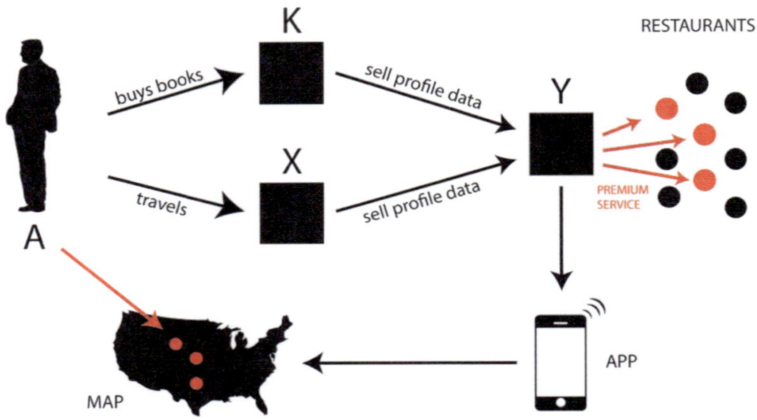

Fig. 4.1 Interacting with restaurants and cities using data

We could choose to analyze these last ones.

As shown in Fig. 4.1, *subject A* is a constant visitor to *Web site K*, which he uses to purchase books and other objects. The Web site records *A*'s preferences and tastes, by observing the characteristics of the product pages he visits online, his purchases, and other parameters (e.g., the time *A* spends on each product's page). The service is able to build a profile for *A* whose quality is proportional to the quantity of time and interactions spent on the Web site.

Another service, *X*, executes similar profiling processes, for travel and tourism.

To access services *K* and *X*, *subject A* has accepted their terms of services[3] of both Web sites, which attest that the profiling information can be shared with third parties.

A certain company, *Y*, is one of these third parties. Both *K* and *X* share (sell) profiling information to *Y*. *Y* integrates the data, obtaining new data, maybe discovering that *subject A* loves oriental cultures and that he frequently visits Japanese restaurants.

Y, maybe, manages a social media site dedicated to food and restaurants and massively uses this kind of data to decide its advertising strategies, providing its advertisers (restaurants, which probably advertise their commercial venue using *Y*) a premium service in which they are granted that their ads will be highlighted for the users whose profiles clearly indicate that they have a preference for the type of food or style of the advertiser's venue and that, thus, may have a higher probability of ending up actually eating at their restaurant.

[3]Terms of Service agreements are legal documents which users of online systems accept when using certain Web sites or services, and which regulate their characteristics, also in relation to the existing laws and regulations. The acceptance may be explicit (e.g., when users are forced to click on an "Accept" button as they are subscribing to the service) or implicit (e.g., when certain Web sites display messages like "By using this Web site, the user declares that he/she has viewed and agreed to the Terms of Service agreement available at [link].").

Y also has a smartphone application, installed on subject *A*'s device.

Subject A finds himself in a new city, which he knows little about. Dinner time comes, and *A* wishes to select a restaurant. To do it, he uses application *Y*. On his mobile phone's screen, he will see a map, on which the Japanese restaurants which have purchased *Y*'s premium service are highlighted and, with all probability, will choose one of them to spend his evening.

This is a very complex (and very frequent) modality for interaction. For example, we can imagine replacing *X* with Amazon,[4] *K* with TripAdvisor[5], and *Y* with Yelp[6].

A user's interaction with social networks and online Web sites generates data and information which is algorithmically recombined in complex ways and used by other services and Web sites which (in this case) generate maps and lists of commercial services in personalized ways which are completely opaque to users.

The user has no way in which he can learn which information he produced has been used as a source of information to generate the map in the *Y* application. Furthermore, these sources may vary each day, according to the evolution of the commercial agreements running between companies.

The result is that *subject A* sees a map which provides a *complex vision of the reality of his surrounding territory*, in which the highlighting of the restaurants is the result of a complex elaboration of his tastes and desires and of how *X*'s and *K*'s algorithms (both invisible for *A*) have been able to identify them.

The map provided by Y represents a complex geography. It is, in part, physical (it effectively represents the position of buildings, streets, and restaurants), emotional and psychological (it derives from the interpretation of A's tastes), and partly economic and financial (being determined by the fact that some restaurants have purchased Y's premium service or not), and we could go on in detailing the list of its technological, algorithmic, statistical, and anthropological characters, and more.

This complex scenario generates a multitude of different ways in which these types of information can be transformed into opportunities for interaction with citizens and other human beings.

Furthermore, the information can be recombined in infinite different ways, to produce infinite different geographies, opportunities for interaction, services, games, artworks, and more.

We find ourselves well beyond Mitchell's "City of Bits" (1996) or of Zook's and Graham's description of "DigiPlace" (2007). We are also well beyond the classical intendment of georeferenced data and of the visions expressed by Aurigi (2008) about the articulation in its digital and physical character.

We find ourselves in a digital Third Space which is not only pervasive and inclusive, but in which information is also not only attached to places, spaces, bodies, and objects, but constantly recombines, remixes, and recontextualizes,

---

[4]Amazon: www.amazon.com.

[5]TripAdvisor: www.tripadvisor.com.

[6]Yelp: www.yelp.com.

creating constantly new geographies which are emotional, linguistic, semantic, relational, or relative to the many patterns which non-human algorithms can glimpse in the ways in which layers emerge from data, information, and knowledge, correlating different spaces, times, and human networks.

4.3 The Third Infoscape, Again

In this vision, the city assumes the shape of a body in perennial mutation, including both architectures and the constant and emergent layering of stories and knowledge which originate from the daily lives of citizens and nature.

At this point, it is possible to go back to our original narrative, to the concept of ubiquitous and emergent knowledge, and to connect it to this vision, to Clément's and Casagrande's vision.

The new types of information, the ones which converge in our perception of the city and, thus, into our interaction with human beings, architectures, spaces, places, and organizations, be them emotional, semantic, linguistic, relational, and relative to the possibility to identify multimodal and multilayered patterns which can be localized anywhere in space-time, whether they are generative or algorithmic, whether they derive from sensors or other interactions, all of these instances of data, information, and knowledge, today, often have digital form and ubiquitous manifestations.

We experience them through smartphones, applications, social networks, interactive services, and systems which are disseminated, distributed, or even pulverized through space and time. Through them, we can interact with the world, express ourselves, collaborate, work, express emotions, consume, study, and entertain ourselves.

As we have seen in the previous chapter, we can define the Third Infoscape precisely here, in the connective tissue which we constantly create, take apart, and create again, in our use of technologies and other data-and-information-generating processes, creating myriads of microhistories which interweave and interconnect nature, buildings, jobs, administrations, emotions, purchases, traversals, bureaucracies, movements, images, and all of the other things which constitute our daily lives: ubiquitous "rumors" (and, in general, expressions), whether they are conscious or unconscious, which interconnect and syncretically map the life of the city.

The Third Infoscape refers to the information and knowledge generated through the myriads of microhistories, through the progressive, emergent, and polyphonic sedimentation of the expressions of the daily lives of city dwellers.

The vision of the new paradigms of interaction with the city is centered on the Third Infoscape, which is, to all effects, an *ubiquitous communicational ecosystem*.

4.4 A Walk in a New City: Dust

Graham (1997) wondered how it could be possible to imagine a real-time city by taking in consideration the ways in which telecommunications reconfigure our notions of time and urban space. This goes in the direction of the definition of a communicational environment, a diffused cloud of sense and meaning which goes beyond the dynamics of screens, and which is not virtual anymore, but impalpable and mental.

This atmosphere is found in the spaces which are in-between, interstitial, and ubiquitous. It is not an idealized representation, but a mobilization of imperceptible urban matter, manifesting itself through pervasive computation which is both automatic and relational.

To all effects, with the development of wireless sensors, of smart dusts,[7] and with the possibility to engage human beings in urban sensing processes, the dimension of virtuality collapses. Heading toward a state which is basically comparable to the one of telepathy (among human beings, human beings and machines, machines and machines, etc.), reconfiguring urban ecologies so that mapping virtuality or physicality would not be needed anymore, and replacing this need with the possibility to create recombinant inventories of the telepathic migration of dusts, of the myriads of pulverized sensors which are disseminated, diffused.

This telepathic form is, thus, a form of invisible communication which describes the ways in which the city talks to itself, circulating messages and reprogramming urban ecologies.

The circulation of messages represents and moves physical shifts and transformations. The city itself moves, as phenomenon and meta-phenomenon. A feedback loop, thus, is created, in which we find ourselves simultaneously immersed and unaware of the—telepathic—exchanges which surround us.

We can imagine information mutating into landscape, delineating an urban space which is not determined by distance and time, but from the transformation of densities and presences as shown in Fig. 4.2.

Gabrys et al. (2010) states that:

«*The wireless city is a space for the production of dust in all its modalities. The city abounds with compressed and errant signals. Yet instead of dissolving urban space, as so many writers suggest, these communication and sensing technologies fill it with signals*».

It is interesting to note, after all, how it is not important that messages arrive to destination and accurately assemble themselves, but that it is possible to understand how these are filtered by noise and dust and the ways in which the most relevant and valued composition comes into being.

[7]Smart dusts are clouds of microscopic sensors which are connected to wireless networks and which can be diffused in the environment. To know more: http://it.wikipedia.org/wiki/Smartdust.

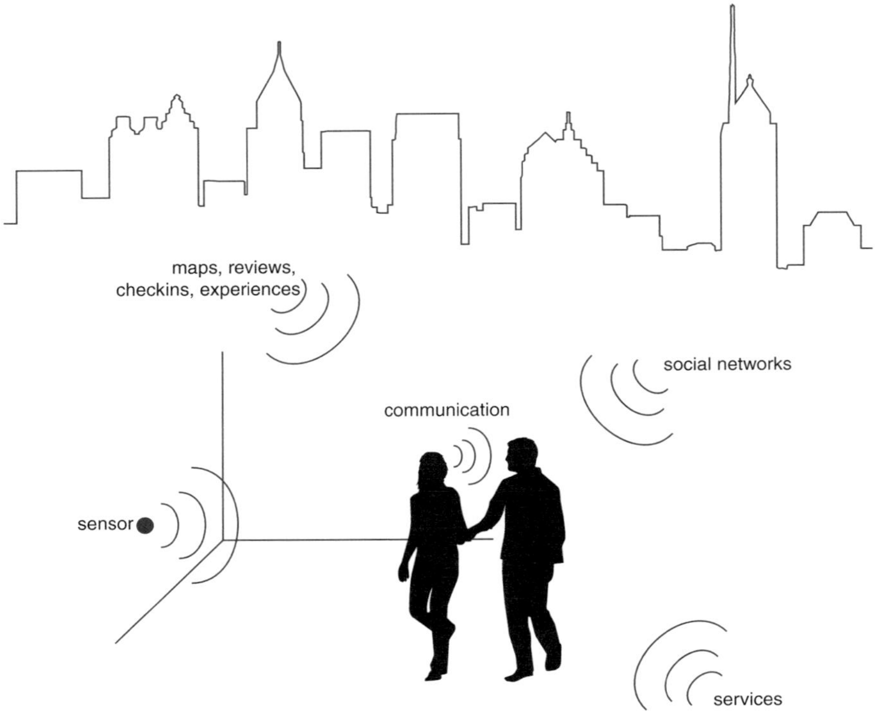

Fig. 4.2 Telepathy and circulation in the city

On this aspect, Gabrys continues:

«*This is the telepathic imperative. Data exists everywhere in excess. In the wireless city, it floats and settles in a hazy surround. Sifting through the modalities of dust to sense and communicate through the urban medium will ultimately require a well tuned telepathic sense*».

In his "Amusing Ourselves to Death," Postman (1985) hypothesizes how the realization of these complex media ecologies would expose us to this type of issue: for the quantity and quality of information; for their structural configuration (in the sense of the type of media and, within it, of the architecture of information); and for their shape (this hypothesis was even more strongly confirmed in Informing Ourselves to Death). The problem, according to Postman, is not in the availability of information, but in the possibility to extract meaning from information.

This type of problem has been highlighted multiple times and defined as information overload, data smog, spam, or under the constructivist form of the attention economy described by Davenport and Beck. The technological solutions at this level are also problematic, at least when they are not oriented toward providing usable, accessible, and inclusive mechanisms for content classification, for filter, and for the expression of their relevance. And—also in these latter cases—the algorithmic dimensions of these processes isolate us from the possibility to comprehend the meaning of information, however remixed. Google's Filter Bubble[8] is a classic example of this phenomenon, with enormous impacts on the reachability of information, on its accessibility, on the rights to expression and information, and on the dangers deriving from the creation of opaque zones in which the mechanisms according to which information are published, hidden, or highlighted are not transparent.

Technical solutions apart—and their corresponding algorithms, systems, interfaces, constantly more advanced to be able to enormous amounts of data, information, and knowledge—the most interesting results come from the transmedia character of information and from their participatory performability.

From the first point of view, following Jenkins'(2006) definitions, transmediality allows us to simplify the extraction of meaning from enormous amounts of information, making its access more immersive and usable: content which is sharable and spreadable and which offers opportunities for mutual interconnection, across different media.

From the second point of view—which becomes important also in evaluating the first one—the problem of overload and the impossibility to extract meaning becomes easier to confront to when messages are freely accessible and performable, and when the ways in which they have been generated is transparent and also accessible, just as the way in which it should be possible to intervene in the flows of their generation, their processing—and remix/recombination—, and their propagation.

In synthesis, this equals to the need to create legibility for the relational graphs related to the generation, processing, and propagation of data, information, and knowledge and to make accessible—in ways which are inclusive—the possibility to intervene, infiltrate, and add in any stage of the process: enabling information to be performable.

Both mechanisms require intellectual property management techniques which are more refined, advanced, and just, if compared to the ones we have available now, from legal and perceptive points of view.

In the discussion about the Planetary Garden,[9] Clément proposes specific questions regarding property which are of fundamental importance in all of this

---

[8]The concept of Filter Bubble, initiated by activist Eli Parser, indicates those algorithms which, by highlighting certain search results—for example on Google—on the basis of the interpretation of what could be more pertinent to us according to our user profile—and on how this is interpreted by online systems—effectively block us from the possibility to access all of the available information, creating, thus, a bubble around us, and avoid us from leaving it. For more information: http://en.wikipedia.org/wiki/Filter_bubble.

[9]For example on Domus Magazine: http://www.domusweb.it/it/recensioni/2013/11/08/lezione_di_giardino.html.

discussion. Ecological dynamics assume the restructuring of the applicability of private property, from the point of view of a mutation of the concept of value and from the point of view of the emergence of what can be described as the dreaming economist and guarantor of a dynamic, mutating, and mutant landscape, not a definitive one:

«Emergent ecosystems could be a source of wealth, but being misunderstood by the system, they are misunderstood by us, as well».

These dynamics attribute a central role to knowledge and to its free accessibility, recombination, and remix, both in terms of usage and in the ones of imagination, education, and sharing.

The redefinition processes for the concept of property (intellectual, in this case)—and on its implications on accessibility, inclusiveness, and usability—become necessary when objects themselves undergo radical transformation.

Complex mutations have already happened to be able to confront with entire market disruptions brought on by the diffusion of mp3, videos, images, and other simple media (simple, in the sense of mono-media). Thus, it can become intuitive how even more radical transitions and transformations will be needed to adapt these concepts to data, information, and knowledge which, now, are of a completely different type.

Services such as Google, Facebook, Twitter, Amazon, and Apple produce data, information, knowledge, and identities which are diffused across different and multiple devices and modalities, interacting in profound ways with the things we know—and that we can know—about the world, its inhabitants, and with the ways in which we experience places, events, monuments, schools, restaurants, workplaces, and a lot more.

Messaging applications reach us ubiquitously.

Devices, sensors, gadgets, wearable technologies, prosthetics, and, soon, entire body and neural extensions interconnect bodies, emotions, health information, movements, gestures, and sensations, exhibiting them on social networks and sharing them—knowing or unknowing, whether we like it or not—with diverse types of services and processes, with human beings and machines.

Furthermore, algorithms create additional dimensions, in which each gesture, movement, or action can be recombined with others and transformed into information and knowledge.

These are territories for which there are no maps, yet. We find ourselves within a gray area in which laws, regulations, and people's perceptions are not defined.

In this scenario, it could be desirable to actuate a shared, open, and inclusive effort to define the Ubiquitous Commons,[10] the commons in the era of ubiquitous technologies.

[10]Ubiquitous Commons: www.ubiquitouscommons.org. A possible definition will be explored in following chapters.

4.5 Public, Private, and Intimate

How is it possible, then, to capture the manifestations of the Third Infoscape in urban environments (and, in general, in a certain territory)?

This is a fundamental question, which gives rise to complex—but feasible—technical answers and complicated theoretical and philosophical critical ones, with profound implications at political, social, and anthropological levels, bringing on potentially radical transformations to concepts such as fundamental rights, privacy, property (both intellectual and physical), and more.

It is, in principle, possible to set up complex technological systems which capture as many manifestations of the Third Infoscape as are expressed in the digital public sphere.

By capturing "public" expressions on social networks and harvesting data from sensors, cameras, transactional systems, services, and more, it is possible to create large repositories of these data and information, potentially—and tendentially—in real time, and to process them for various purposes to understand: where they are generated; what relations they express between people, objects, places, organizations, actions, events, etc.; what emotions they express, for example by performing natural language analysis processes (Iaconesi and Persico 2015); or what topics and issues they deal with. Data can, then, be integrated in multiple ways, wherever appropriate and possible, ensuring that the integrations generate meaning. In this way, it could be possible to obtain a massive availability of the manifestations of the Third Infoscape in the city.

It would seem, thus possible, to set up such systems, start the data and information harvesting processes, analyze the resulting data in a number of ways, and proceed to use the Third Infoscape to bring emergent, active, responsible, participatory, and inclusive performance into cities, through a number of services, educational processes, spaces for citizen interaction, and collaborations between institutions, organizations, enterprises, grassroots initiatives, and single individuals.

But, instead, this potential technical feasibility hides a large number of fundamental issues that must be confronted with:

- the definition of the digital public (and private) sphere and, as a consequence,
- surveillance and privacy;
- inclusiveness, access;
- divides and literacy;
- transparency and the algorithmic domains; and
- legal issues.

We will deal in detail with most of these issues in the final chapters of this book (namely Chaps. 11, 13, and 14), where we will analyze the social, political, anthropological, and psychological implications of such systems and processes—after having described them in some detail—and describe possible actions and organization models—for citizens and institutions—which can lead them in positive, constructive, and inclusive directions.

But, in the meanwhile, it may be useful to ground a short appraisal of these issues, so that we will be able to keep them in mind as the rest of text goes on.

Almost all of these issues are interconnected with each other, and they deal with the need to reconsider definitions which we have, historically, accepted and given for granted, starting with the definition of public, private, and intimate spheres.

Past definitions of public spheres (and of private and intimate ones) need to be reassessed, in order to evaluate how (and if) they are still applicable to the current world (and, in tendency, to the world of the future) and how we would like them to be adjusted.

In the world of ubiquitous technologies, pervading our homes, offices, schools, streets, elevators, stores, air, and even our bodies and minds (just think about prosthetics, wearables, biotechnologies, and social networks), the definitions of public space coming from scholars such as Habermas, Lefebvre, Arendt, Mitchell and others are just not adapt in representing the current scenario.

Due to the technical, technological, and legal (not to mention political) scenario, it is currently impossible, on the one hand, to understand the architecture of digital spaces and the implications in terms of freedom of expression, assembly, performance, privacy, and surveillance and, in general, to understand how the data, information, and knowledge we produce is used, and how, on the other hand, it is also currently impossible to express and control how this data/information/knowledge should be used.

Put in simple terms, it is currently not possible to understand what "public," "private," and "intimate" are. Even simple actions like configuring our privacy settings on major social networks are misleading in this: while we may think that we are actually deciding who will be able to read (or not) the information we publish, what we are doing is, in fact, making this decision in relation to a limited number of use cases. The social networking service provider will be able to access our private messages for technical, administrative, and business motivations. Algorithms—of which we know nothing about—will be able to process the messages or relevancy to a number of issues, so that they will profile us for anything which goes from advertising purposes, to national security, whatever other imaginable reason. More algorithms will process what algorithms have processed, producing meta-metadata, feedback loops, system behavioral changes, implications on our behaviors, and more.

If this situation is already complex with social networks, it becomes impossible to grasp when we arrive at scenarios involving the Internet of Things, wearable technologies, and biotechnological implants.

For average users (and not-so-average, including hackers and technology experts), it is currently impossible to understand what data they actually produce, how/when/where/why it will be used, and, also, to express (and, eventually, enforce) how they wish for it to be used.

This scheme leads, as said, to the impossibility to determine boundaries (however mutating and dynamic) between public, private, and intimate spheres in ways which are certain and supported by evidence and to a generalized state of surveillance—whether it is real or only perceived—due to legal apparatuses, the opacity, and closedness of technological systems, and to the yet unexplored (and potentially unexplorable) world of profiling algorithms for people, places, objects, events, flows of communication, and interaction.

This situation is rendered worse by literacy and access issues. Not everyone is "connected," in the many ways in which this word may manifest its meanings: from the case of someone who is digitally illiterate (maybe because of their age, or for the place in which they were born or in which they live), to the other extreme in which someone might desire to not be connected, and everything that falls somewhere in between these two extremes, in a continuous gradient of possibilities which involve cultures, places, ages, desires, expectations, personal histories, and more.

This situation has multiple types of effects, ranging from the possibility for digital spaces to actually be representative of the expressions of large (if not all) parts of society; to the one of understanding how to address citizen participation and activation processes in inclusive ways; to the transformations of the education system(s), in order to diminishing divides and creating more access; to the actual possibility to decide whether to be connected or not, even in spatially and temporally localized scenarios; and to the actual possibility to defend and protect fundamental rights for human beings.

4.6 Human Ecosystems and the Ubiquitous Commons

To address all of these issues in our research and practice, we have combined a set of different approaches, producing a series of tools and methods which incorporate technologies, legal aspects, as well as cultural and educational actions whose intent is to stimulate the imagination of wide parts of the population in order to induce a possibilistic attitude and a performative and participatory state onto which it is possible to build new ideas and operative modalities. Arts, design, and, in general, creativity play an important role in this process, as ways in which it is possible to bring people together around new aesthetics, engaging experiences and imaginative narratives whose purpose is to produce novel, yet uncharted mental spaces in which it becomes possible to move and create, collaboratively.

The two main assets for this strategy are called *Human Ecosystems* and *Ubiquitous Commons*.

Human Ecosystems[11] (HE), the first, is a combination of a technological system, a novel type of museum, and an inclusive education process. HE is a city-based project, which has been started in Rome in 2013 and instanced multiple times in cities across the world such as Sao Paulo, Malmö, Berlin, New Haven, Toronto, Montreal, Lecce, Bari, Bologna, Turin, Milan, and more, in different ways and at different scales. All the case studies in the book are only some of the examples in which HE has been used.

[11]Human Ecosystems: www.human-ecosystems.com.

When HE[12] is started in a city, a technological system is set up to capture all of the city dwellers' public expressions on major social networks[13]: All of these data and information are processed using a number of techniques: various forms of natural language analysis to understand what people are discussing, around which topics and addressing which issues, and to gain insights about what emotions they are expressing; multiple types of geo-referencing techniques, to understand for which places in the city the data and information are relevant (is the person in the place, going there, talking about the place…?); network analysis to understand how the captured content expresses relationships among citizens, groups, organizations, institutions, or companies, and according to what models and patterns information, knowledge, emotions, opinions and points of view flow and spread in the city; and relational analysis to interpret these and other factors to gain better understandings about the relationships that these expressions and their impacts describe within the city and interconnecting the city with other places.

These data are released as a real-time source of Open Data, constituting a commons which can be accessed, used, managed, and explored by all types of stakeholders, ranging from researchers, public administrators, designers, artists, students, citizens, and entrepreneurs. Access is regulated according to a reciprocity licensing scheme, which comes under various forms depending on national and international regulations, by which non-commercial and fair use is void of any restrictions and commercial usage must imply the generation of some form of proportional reciprocity, some positive outcome for the city.

The data are used as the engine for the implementation of the Real-Time Museum of the City (RTMC). The RTMC (an example of which is shown in Fig. 4.3) is an *iconic* and suggestive location in the city whose objective is to establish the aesthetic, cultural, and perceptive presence of the Third Infoscape in the city. It comes under the form of a public space in which the Third Infoscape can be experienced and performed through a series of interactive information visualizations, installations, and immersive environments. By bringing the Third Infoscape to fruition in powerful, engaging, and emotional experiences, it aims to achieve the construction of an extended mental space about what the city is and can be, with all of its flows of communication, interaction, emotion, opinion, rumors, desires, expectations, and visions. The RTMC is participatory by definition, as its content is created by the polyphonic manifestations of the myriads of microhistories of the city, generated by all of its data and information subjects, whether they are citizens, entrepreneurs, organizations, sensors, and cameras.

The museum also brings forward a *critical* vision, in multiple ways.

For example, one of the experiences allows visitors to find themselves in the Third Infoscape. By entering their username on major social networks, people can

---

[12]To learn more about the technical and technological details of Human Ecosystems, it is possible to look at the Web site for the project or to read Iaconesi and Persico in "Urban Acupuncture in the Era of Ubiquitous Media" (2014).

[13]Currently, Facebook, Twitter, Instagram, Foursquare, Flickr.

Fig. 4.3 The Real-Time Museum of the city in Sao Paulo, Brazil

find themselves in the museum, as content, data, and information which they have generated and which has been captured in the harvesting process become highlighted on the visualizations. The things they discuss about; the relationships which they manifest; their emotions; their positioning within the human networks in the city, including the indication of the agreements, disagreements, affinities, and more.

This is a potentially shocking experience. "What am I doing exhibited in a Museum?" "Why am I in the museum of the city?" "Who asked for my authorization for including me in the museum?" While some people welcome this feature of the museum in positive, excited, and possibilistic ways, these are just a few of the most critical questions which are asked by all sorts of citizens, tourists, and city dwellers visiting the museum. This highlights the worries, misunderstandings, and perceptive faults which surround the effects of digital technologies on our societies, for example in terms of surveillance and privacy. While the Human Ecosystems only captures "public" data and information (the ones for which it is, e.g., not even necessary to be logged in major social network for them to be visible and accessible; the ones for which if you are publishing them in that way, it is as if you were publishing them on a major newspaper, such as the New York Times), there is still no widely shared comprehension about the extent to which these data and information are available for usage by people, organizations, algorithms, and companies. The interface politics do not help in that, by establishing interactive environments in which little space is given to the possibility to reflect on the critical implications of what users publish. Terms of Service agreements also do not help, by establishing contracts which very few people read and even less people understand (or are able to understand). These facts have enormous implications for our privacy and our possibility and freedom to communicate and express, on our professional lives, our families of any kind, and more.

And this is only one of the critical visions which can be and which are generated through the RTMC and for which the museum is designed to provide information, knowledge, and tools (e.g., to protect oneself, or to establish just who will be able to access the information we generate).

The RTMC also hosts a wide and inclusive education process. This iconic public space becomes the place in which anyone can learn how to use (or how to protect themselves from) the Third Infoscape of their city. Games, courses, and workshops designed for children show them how to use it to learn more about their city, to find new friends, or to understand the different cultures which live in their urban environment. Students can learn how to use it for their school projects. Adults can learn how to use it as a driver for their civic life, to organize themselves, to collaborate with other citizens on the issues they care about, and to establish participated decision-making processes. Artists can learn how to use it to create generative artworks. Activists apprehend how to create public engagement on important themes. Elderly people can understand how to become more engaged in the life of the city, or how to grasp what they are missing in terms of relations, events, and discussion, because of their lack of digital literacy. Researchers learn how to use it to perform anthropological, sociologic, psychological, and other forms of research. Designers learn how to use it to create novel services, communication campaigns, and generative objects whose form depends on data, or the practices of data and information-driven collaboration and cocreation. Public administrators can learn how to use it to establish shared, participatory decision-making processes, or collaborative policy-making efforts. Everyone learns how to deal with the ethical issues included in all of these practices, how to actively engage citizens, how to ask for permission, and how to protect one self from misuse and malice.

All of this happens in this public space, in transparent ways, with the inclusion of critical approaches, with everyone involved learning and being literally surrounded, immersed in the Third Infoscape of their city, as the workshops and lessons happen directly inside the museum, the effects of experiments and exercises being immediately visible on the large information visualizations, and, eventually, with the possibility for them to be included in the museum itself (e.g., as it happened for the RTMC in Sao Paulo, in Brazil, with the explorations about the water crisis of the city, as we will see in Chap. 12).

This combination of open, accessible tools; aesthetic, perceptive, and cultural interventions; education processes, together with the establishment of a public space to engage large parts of the population in inclusive ways, to generate discussion and civic performance, and the wide usage of critical approaches, contribute to a transformative process in which the Third Infoscape becomes perceived as a living part of the urban environment, in which life and participation can (and should) take place.

The other framework which we use in our research is composed of the legal and technological tools which are described as the Ubiquitous Common (UC).

The UC address many of the main critical issues which regard our lives as data subjects.

As stated earlier, in the current scenario, people and organizations generate huge amounts of data but:

- They cannot know in exact ways what data they generate;
- They cannot know what data and information are derived from the data and information they generate;
- They cannot know how these data and information are used, and who will be able to use it; and
- They cannot even express (and enforce) how they would wish their data were used, and by whom, at which conditions.

UC aims at solving some of these issues and, in the meanwhile, to create a high-quality relational environment in the common and participatory management of data. These solutions are schematized in Fig. 4.4, which shows the scenario before Ubiquitous Commons, and in Fig. 4.5, which shows the scenario after Ubiquitous Commons.

Ubiquitous Commons is a shared global research effort dedicated to understanding the transformation of data, information, and knowledge in the age of ubiquitous technologies and networks. The project aims to create a legal, technological, and philosophical toolkit transforming the ownership of data into a relational concept, in which individuals and communities can actively cooperate in the attribution/definition of rights/duties of access to the data through digital interactions (from social networks, to apps, sensors, wearable technologies, devices, Internet of Things, CCTV cameras, security and surveillance schemes, algorithms and processes of various types, and so on, consciously or unconsciously).

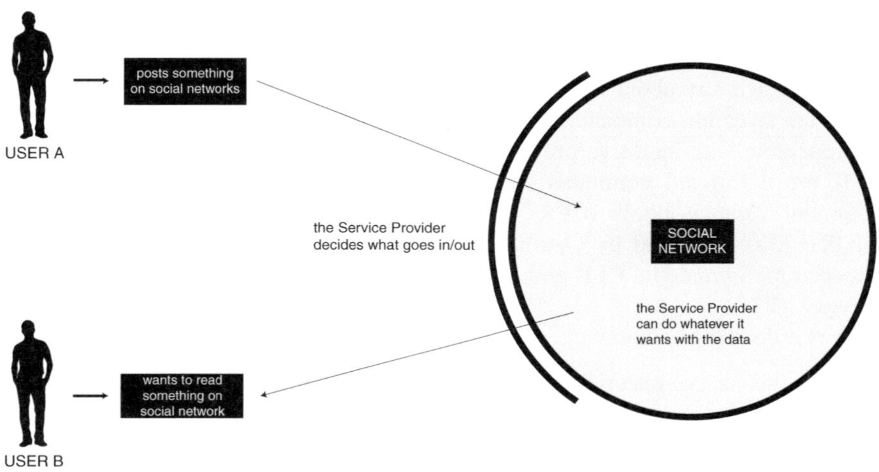

Fig. 4.4 The scenario before Ubiquitous Commons

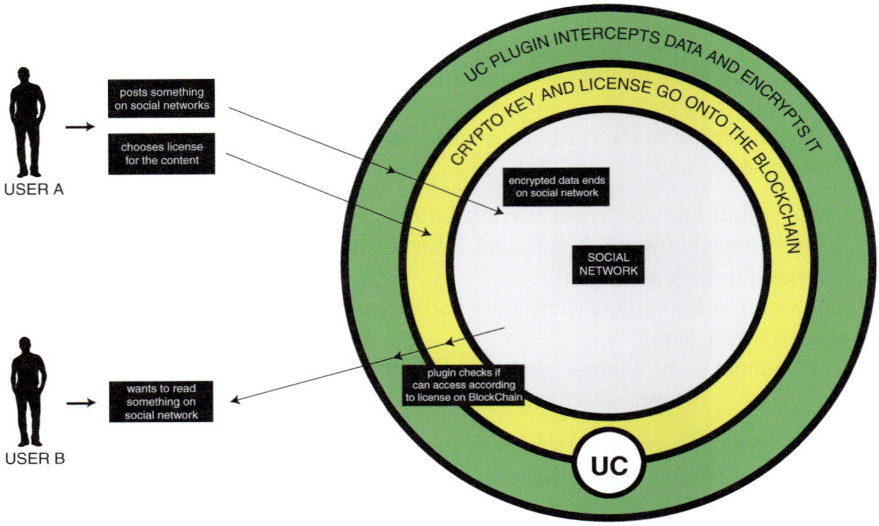

Fig. 4.5 The scenario after Ubiquitous Commons

The first prototype of the technological toolkit is dedicated to social networks and Web/online services. It is a browser plug-in combining encryption, p2p networks (the Block Chain, the p2p network behind BitCoin, the most popular digital currency), and a mechanism for "user generated license." Once installed, the plug-in intercepts the content we are to publish, encrypts it, allows you to generate and apply the desired license, and, only then, sends it to the service. Both the decryption keys and the licenses are distributed on Block Chain, meaning that people can decide by whom and under what conditions the content may be used. The result is a cooperative, relational, and totally p2p mechanism in which individuals, communities, institutions, companies, and organizations—beyond the unspoken "law of Tos" (Terms of Services) established by the service providers—can have their say about how their data are used, creating new types of licenses: civic, for research, commercial, for a fee, or entirely personal, based on an open, interoperable, and inclusive protocol.

If the traditional commons depict the strong relation between the material common resource pools (CPRs) and the high-quality relational environments (HQREs), as described by Ostrom (1990), the UC highlights the strong relation between the immaterial CPR and the HQRE created by establishing a peer-to-peer network (P2P).

It is a double transition:

- from the scarce, material resources to the abundant, and immaterial resources;
- from the physical relational environment, to a relational environment that can be either physical, digital, or hybrid, and which is expressed through a P2P network enacted through person-to-person relationships, social networks, Internet of

Things (IoT), sensors, network connected devices, databases, and processes, using the Ubiquitous Commons and keeping the quality standards.

It is here, in the P2P network, that the purpose of social conscience, imaginary, and sense of responsibility—which are typical of the HQRE—are enacted, and here is the place where they form the feedback loop to P2P network itself, constituting a second-order cybernetic system.

As we will see in later chapters (mainly in Chap. 13), this will be the fundamental step in designing novel organizational forms and strategies, focused on peer-to-peer networks, interconnected, and interrelated with each other, with a sense of meaning, purpose, and responsibility, living together within in a Relational Ecosystem according to a logic of coexistence, instead of consensus.

4.7 An Interview with Carlo Ratti

An architect and engineer by training, Professor Carlo Ratti directs the MIT Senseable City Lab and is also a founding partner of the international design and innovation office Carlo Ratti Associati. He graduated from the Politecnico di Torino and the École Nationale des Ponts et Chaussées in Paris, and later earned his M.Phil and Ph.D. at the University of Cambridge, UK. As well as being a regular contributor to Project Syndicate, he has written for international media including New York Times, Washington Post, Financial Times, Scientific American, BBC, Il Sole 24 Ore, Corriere della Sera, Domus. His work has been exhibited worldwide at venues such as the Venice Biennale, the Design Museum Barcelona, the Science Museum in London, MAXXI in Rome, and MoMA in New York City.

1. **To "Read" the city. This implies the possibility to acquire data and information, and to interpret it, generating understandings and meanings. The term "read", in English, originates from "raten", which means "to guess, to advise". This implies the fact that the data and information collection process is just as important as the interpretation process, and that different "readings" can generate different interpretations.**

How, in your experience, are we (scientists, researchers, administrators, planners, designers, citizens) getting ready to confront with the massive amounts of data and information which is and will be available to us, to allow for emergence of valuable interpretations of our cities, enabling difference and, at the same time, being able to make sense of this abundance?

Collecting data are actually the starting point of city planning. Over a century ago, French geographer Élisée Reclus claimed that all planning exercise should start from surveying and the collection of data. The only difference today is that we have access to a staggering amount of data, collected in real time. So there is nothing new in today's approach—just an immensely larger amount of information. This very condition makes it more difficult to analyze and interpret the data. At the same time, however, increased computing power is helping with this task.

Regarding the possibility of different interpretations of the data, it has always been like this. Data can be appropriated by partisan groups and misused. The nineteenth century British Prime Minister Benjamin Disraeli used to say that "there are three types of lies: lies, damned lies and statistics."

(2) **Data and information are ubiquitous. They are generated in implicit and explicit ways; in conscious and unconscious ways; in ways that are inspectable and in ones which are completely opaque and inaccessible; in ways which are acceptable and in open breach of people's rights and respect for privacy, intimacy. On the one hand: how do you perceive the rising conflicts which are and are about to raise regarding the massive data and information harvesting processes which we experience today? What are the dangers to look-out for? What are the things you believe will shape our concepts of privacy, intimacy, publicness, intellectual property?**

Before the digital revolution of the late twentieth century, Italian writer Italo Calvino imagined a peculiar mnemonic condition in a short story titled "The Memory of the World." In it, Calvino imagines a society—a gently terminal dystopia—in which every detail and every moment of humanity and of the world is recorded for posterity—blending scholarly knowledge with the mundane details of life. Needless to say that such effort leads to drama. The end of the story is a mix of intrigue and paradox, and puts an uncanny spotlight on the importance of discussing this very topic: who has access to the collected data? How does this turn into power? And finally, how will humanity remember itself?

The questions above are very important today, and it is crucial that we all participate in an open discussion about them. That's why we already organized twice a forum at MIT called "Engaging data" (http://senseable.mit.edu/engagingdata2013/) —inviting academics, privacy advocates, the US Administration, companies that collect Big Data, etc.

(3) **On the other hand: these data and information are, to all effect, in the hands of only a few operators. Open Data processes are not enough, if they are not paired to wide literacy and education processes, and by the development of a popular imagination and desire about what you can do with data (as a citizen, for example), and the creation of the high quality relational environment which is needed for citizens to do things together, embracing collaboration and peer-to-peer action. In the meanwhile, information visualisation are starting to become a "new TV", proposing packaged interpretations which are starting to enter the domains of the spectacle and of entertainment. How, in your opinion, human beings are and will be able to benefit directly and autonomously from all of this ubiquitous data?**

We believe in a ground-up approach to creating cities in which people can be the main agents of change. Open data and visualizations can be very effective to this end. Let me mention one example.

In our Trash Track project in Seattle, i.e., we added tags to trash and then followed trash as it moves through the city's sanitation system. Out of the many things we learned is that the simple sharing of information can promote behavioral change. People involved in the project would be able to follow their trash and this led some of them to make different decisions. One person told us: "I used to drink water in plastic bottles every day and throw them away and stop thinking about them. Now after the project I know that these bottles go a few miles from home to a landfill and will stay there forever. As a result, I stopped drinking water in plastic bottles."

(4) **Algorithms, complexity and accessibility. The algorithms which regulate so many aspects of our daily lives—from the stock exchange, to the way our heating system works, to what we see on our apps or social networks, to the ways in which our insurance risk is calculated, etc.—are becoming progressively more complex. So much that we really run the risk of not understanding their implications anymore. Machine learning, neural networks, artificial intelligences really do not speak the language of human beings, even if human beings design them and use their results. This type of phenomenon (complexity getting out of hand) has already happened with other systems (e.g. the Stock Exchange, for example with the Flash Crash of 2010 and similar phenomena). In your opinion, how will we, as human beings, be able to deal with the enormous levels of complexity introduced by computational research and practices? How will we be able to really understand what happens in systems of this kind, and to interpret the results which they provide us (for example: the Google's filter bubble allows us to see "what we like", also creating a bubble around us, making information unreachable)? How are we dealing with the social, political, philosophical and psychological implications of these phenomena?**

I do not think that we will see a fight between the natural and the artificial world, which is behind many "singularity" scares. I believe more in cyber-biological systems, mixing the strength of humans and machines.

(5) **Researchers such as Graham and Gabrys speak about data and information in cities in ways which describe the formation of a novel mental space and landscape, which become cognitively accessible to us as access techniques and interfaces. Gabrys arrives at provocatively defining the possibility to be "telepathically urban", alluding at the cognitive connection with data (and with other people, through data and information) which we are now able to establish. This, of course, has enormous implications at cognitive levels, for example in learning how to deal with information overload, or with irrelevance, or to understand how to set priorities, work and study effectively, establish meaningful relations, etc. What are the tools and techniques which, in your opinion, we will learn how to use (as citizens), to deal with these novel mental spaces? On the one hand to deal with abundance and, on the other hand, to be able to be able to maintain our rights, autonomy and freedoms?**

I remember a dinner with James Flynn, who discovered the Flynn effect, a few years ago. He told me that one of the most plausible explanations for the raising of IQ over time is urbanization—which exposes to us to richer environments. So I do not see a fundamental difference with what is happening today.

References

Aurigi A, De Cindio F (2008) Augmented urban spaces: articulating the physical and electronic city. Illustrated. Ashgate Publishing, Reading
Cullen G (1966) The scanner. Alcan Industries Ltd, London
Cullen G (1968) Notation: the observant Layman's guide to his environment. Alcan Industries Ltd, London
Cullen G (1971) The concise townscape. Architectural Press, London
Gabrys J, Boutros A, Straw W (eds) (2010) Telepathically urban. In: Circulation and the city: essays on urban culture. McGill-Queen's University Press, Montreal, pp 48–63
Graham S (1997) Cities in the real-time age: the paradigm challenge of telecommunications to the conception and planning or urban space. Environ Plann A 29(1):105–127
Iaconesi S, Persico O (2015) Visualising emotional landmarks in cities. In: 2014 18th international conference on proceedings of information visualisation (IV), pp 408–413
Jenkins H (2006) Convergence culture: where old and new media collide. NYU Press, New York
Lynch K (1960) The image of the city. MIT Press, Cambridge
Mitchell WJ (1996) City of bits: space, place, and the infobahn. MIT Press, Cambridge
Ostrom E (1990) Governing the Commons: the evolution of institutions for collective actions. Cambridge University Press, Cambridge.
Postman N (1985) Amusing ourselves to death: public discourse in the age of show business. Penguin, New York
Rossi A (1966) L'Architettura Della Città. Marsilio, Padova
Venturi R (1977) Learning from Las Vegas: the forgotten symbolism of architectural form. Revised edition. MIT Press, Boston
Venturi R (2002) Complexity and contradiction in architecture, 2nd edn. The Museum of Modern Art, New York
Zook MA, Graham M (2007) Mapping DigiPlace: geocoded Internet data and the representation of place. Environ Plann B Plann Des 34(3):466–482

Chapter 5
A Case Study: Turin

Abstract There are multiple cultures in the city of Turin. In one year (2011) of observation of the city of Turin through social networks, we have been able to detect the many languages spoken in the city, as evidence of the multiple cultures which inhabit it. In places, crossings, emotions, communication, opinions, the city is different according to the cultural eyes from which you look at it. Building upon Kevin Lynch's concepts of imageability and readability, the city shows its multiple lives and the ways in which they relate, interact, move, interconnect, and divide.

The dataset for the Turin case study is available at http://www.human-ecosystems.com/DAP/DigitalUrbanAcupuncture-CaseStudy-Turin-dataset.zip.

In September 2011, we were called at the Piemonte Share Festival,[1] held in the Natural Sciences Museum of the Piedmont Regional Administration in Italy, to perform an observation of the real-time life of the city of Turin.

The focus was to try to explore whether it was possible to use the public content which is constantly produced by city dwellers to gather insights about the city's cultural, emotional, touristic, and daily life.

For that occasion, we used a seminal version of what would become the Human Ecosystems technologies—which would be officially born only in 2013— named as the project: Versus. At that time, we had already used the current version of a series of technologies to perform wide-scale observations of cities through their citizens' manifestations on social networks. Possibly the most interesting previous experiment on this had been the "Rome vs Riots"[2] project, in which we observed public unrest in the city (protests, violence, and security issues) and through which we had observed, on October 15, 2011, the citywide violent riot which took place in occasion of the Occupy Wall Street movement, in one of its international actions.

---

[1]Piemonte Share Festival: www.toshare.it.

[2]More information about the Versus project can be found here: http://www.artisopensource.net/projects/versus-the-realtime-lives-of-cities.html

A video showing some of the results of the observation of the riot in Rome on October 15, 2011, is visible here: https://www.youtube.com/watch?v=eVvIGQ3Q0Jg.

S. Iaconesi and O. Persico, *Digital Urban Acupuncture*
DOI 10.1007/978-3-319-43403-2_5

We started an observation of the city of Turin which was almost one year long, and we used all of the captured information—together with some other sources of data and information, to be able to put the one harvested from social networks into context—to try and understand the multicultural life of the city.

Table 5.1 illustrates some statistics about the data captured in Turin.

The graph in Fig. 5.1 shows the temporal evolution of these quantities.

Figure 5.2 shows the concentrations of messages in various parts of the city at different times.

We started the analysis of the data by trying to understand some of the characteristics of specific places in the city.

For example, we gathered all of the content, messages, images, and videos published at Piazza della Repubblica, in Porta Palazzo, to classify them according to the evidence which we were able to discern from them about how people used the square. The Porta has a complex life during its daily cycle. On weekdays, in the morning, it hosts the largest European open air market, with thousands of sellers, stable, mobile

Table 5.1 Some statistics about the data we captured

Number of days	250
Number of harvested contents	6,972,422
Number of users tracked in their public expressions	792,174
Of these, number of "citizens" (users whose contents were generated from the territory of the city of Turin for at least 70 % of the time)	201,496
Number of "tourists" (users whose content was generated from the territory of the city of Turin at most 20 % of the time)	138,994

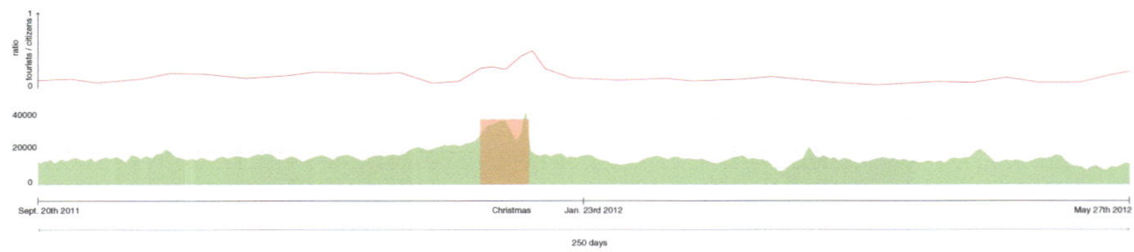

Fig. 5.1 The evolution in time of the data capture in the city of Turin

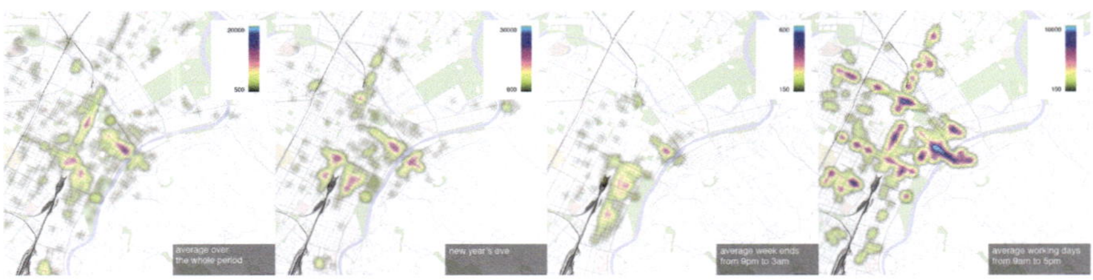

Fig. 5.2 Concentrations of messages in various parts of the city (average messages per day)

and nomadic. It is a natural melting pot, in which people from multiple countries and with very different personal histories come together for a variety of reasons, to shop for the beautiful vegetables on sale in the market, to meet people from their originary country, to perform businesses from the more legal ones to the most illegal ones, such as drug dealing and trafficking. At night, it becomes a peculiar car parking space, with the surrounding houses and the quiet neighborhood life.

We used the harvested content to search for evidence of all of these many usages of the square.

The result of this first type of experiment was a continuously evolving map, and a photograph/text/video exhibit shown at the Natural Sciences Museum.

An example of the map can is shown in Fig. 5.3.

The image shows an aerial view of the octagonal piazza and, overlaid on top of it, the places in which people shared texts, images, and video about certain topics. In the image, for example, are shown people addressing the beautiful market, romantic encounters, religious expressions, and drugs-related issues. Whether it was through people posting images and videos of the beautiful and colorful vegetables, or market stands; describing romantic encounters under the *portici*; or witnessing the unexpected islam worshippers stopping in the middle of the market to pray, and deciding to portray it; or worried citizens of the neighborhood taking pictures of drug dealing happening in unusual hours of the day, under the portici or in the marketplace, at hours of distance from its business life; or in the many other

Fig. 5.3 A day in Porta Palazzo

ways in which people decided to spontaneously document any of these issues and themes, starting from what they saw, heard, and experienced in their daily life.

An interesting scenario started manifesting itself through this simple experiment. It was possible to notice how the best perspectives to take a picture from were distributed along the space; where the most romantic spots in the square are; which are the most culturally significant places for different cultures, because people meet there, or wait, or pray; or where the more secluded, unseen spaces are, so much that people decide that it is safe to deal drugs or engage in prostitution there. Perspectives which, if analyzed, could be extremely meaningful for administrations, planning and for the active, civic, cultural life of the city.

We then headed to extend the reach and scope of these considerations, by trying to gather deeper understandings about where the diverse cultures present in the city express themselves during their daily lives. We started from languages. Each day multiple languages are expressed in the city, which are among the most interesting expressions of the multicultural life of the urban context: diverse sounds, signs, characters, and words not only provide an incredible richness, but also reflect on gestures, forms of expression and social interactions, so much that simple phenomena such as "counting with our hands and fingers" are drastically changing through the effects of this kind of environment: if in Italy it is common to count (e.g., from one to five) using, in progression, the thumb, the index, and through to the pinkie finger, other cultures count differently, for example Middle Easterners start from their little finger, Americans start from their index, Japanese start from an open hand and progressively tuck fingers in, and so on. The exposition and interaction within multicultural environments is changing these gestures and (and the words and concepts which go with them) and creating new ones.

How do different people speaking different languages live the city? Where/when do they interact, cross, or interpret each other? What can this type of information tell us?

Thus, we started classifying the different languages, for how they manifested themselves in the city, and for how they related, interacted or tried to create flows of information and knowledge (e.g., through translation) with each other.

As it is shown in Fig. 5.4, different languages have different places. This changes even more if we take into account the topics which they are addressing, or

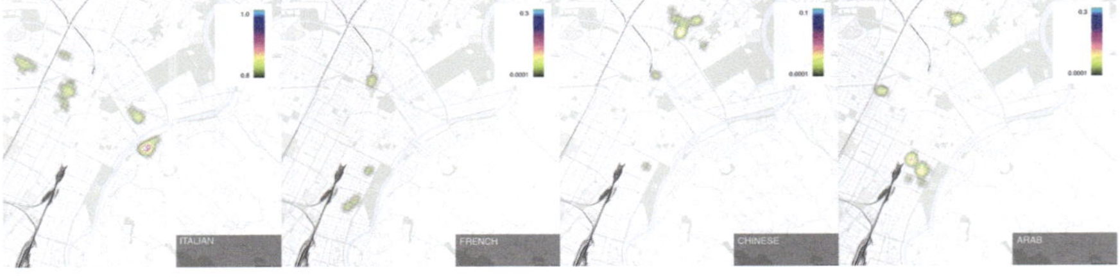

Fig. 5.4 The distribution of different languages in the city of Turin (the gradients show where in the city are the maximum ratios of the expression in the specific language, over the ones in all languages)

the ways in which different languages traverse the city throughout the times of the day, waking up in the morning in a certain place, going to work in a different one, conducting a social life and entertaining themselves on yet another one, and finally going back to sleep. Where do they cross each other? How do they interact? These and more are questions which could be explored using these kinds of tools.

For example, in the image, the "Italian" part of the map shows in which neighborhoods communication happens almost completely in Italian. These are, almost constantly, expensive neighborhoods, like the homes across the river near Piazza Grande Madre di Dio. On the other hand, French and Arab maps show some focal nodes in the city where north African immigration takes interesting forms, like the Barriera Milanese neighborhood, up north, in which many people from Morocco and other French- and Arabic-speaking countries have their homes, or the San Salvario neighborhood, where there is a strong presence of commercial venues and cultural centers from these countries. And as in the case of the Chinese expressions, having impacts on urban communication in places like Porta Palazzo, one of the main melting pots in the city and where there is also a concentration of commercial venues of Chinese origin, or, again, in Barriera Milanese, where are their homes.

Also interesting is the recurrence of user accounts which systematically speak two or more languages (e.g., Chinese and Italian, Arabic and Italian, French and Italian), which, in many cases, characterize the younger generations of immigrants (from the third generation onward), who become perfectly integrated in the city, and who can act as potential bridges among communities (we will learn more about this in Chap. 8).

Having acquired some insights about how different cultures manifested themselves in the city, we proceeded to analyze its emotional expressions.

Emotions are an important part of the life of the city, contributing to describing and activating anything from how well we feel in the different places of the city, to helping us understand how we take decisions, consume, access services, protest, and more. Individuals' emotions; emotions which propagate, as a contagion, traversing relations and networks; induced emotions (through advertising, communication, political campaigns, etc.); and more. Emotions and their combinations determine many of the things which take place in cities. This is why we have focused so much on them, for example, through our concept of the Emotional Landmark[3] (Iaconesi and Persico 2015).

---

[3]Emotional Landmarks are places in which specific emotional expressions are systematically produced by multiple city dwellers, depending on time, culture, context, social group, and other parameters.

In this action on the city of Turin, we tried to understand how certain emotions were expressed in different places, times, and situations.

We used natural language analysis techniques, explained in technical detail in our "Urban Acupuncture in the Era of Ubiquitous Media"[4] to perform what is called Latent Semantic Analysis, through which we checked the texts of the contents published on social networks against concepts and expressive modalities which were relevant for certain emotions. In this, we analyzed texts in 29 languages, against five emotions (love, hate, fear, joy, and surprise).

For each of these emotions, we produced a study whose outputs came under the form of printed information visualizations, real-time generated videos, and generative sculptures, to be produced periodically with the idea of creating a physical archive of the emotional expressions of the city, and prefiguring the concept of the Real-Time Museum of the City.

The most successful emotion, appealing to the excited appraisals of the visitors of the exhibit and of the scientific community, was love.

How does "love" manifest itself in the city? Love between human beings. Love for a culture. Love for art or design. Love for our dogs and pets. Love for the city or for a specific location in the city. Love for our car or for our football team. These were just a few types of expressions which were produced by social network users in the city.[5]

The generative videos shown in Fig. 5.5 were created by interrogating the databases of the harvested content and using the geo-referenced expressions of love in the city to create a parametric surface floating above the map, to show their quantities, as can be seen in the image above, together with the time stamp of the time in which it was captured. Visitors of the exhibit could scroll back and forward in time to appreciate the variation of the quantities of expressions in various moments in time.

The production of the generative sculptures followed the same process. Data from a certain time frame was aggregated (Fig. 5.6 shows the sculptures resulting from one month of aggregated data) and used to create a three-dimensional parametric surface whose elements were geo-referenced (each control point of the surface corresponded to a grid of points on the map): The higher the value of messages expressing love in the area, the higher would be the quota of the control point on the surface. The 3D models generated in this way were converted to formats which could be imported in software packages which could control digital fabrication processes, and they were laser cut, in one case, and 3D printed, in the

[4]All of the software tools and algorithms composing Human Ecosystems are open source and can be found on the popular software code sharing platform GitHub, at the address https://github.com/xdxdVSxdxd/HumanEcosystems.

[5]The video of a few moments in the real-time life of love in the city of Turin. https://www.youtube.com/watch?v=gCtsQEdwTJA.

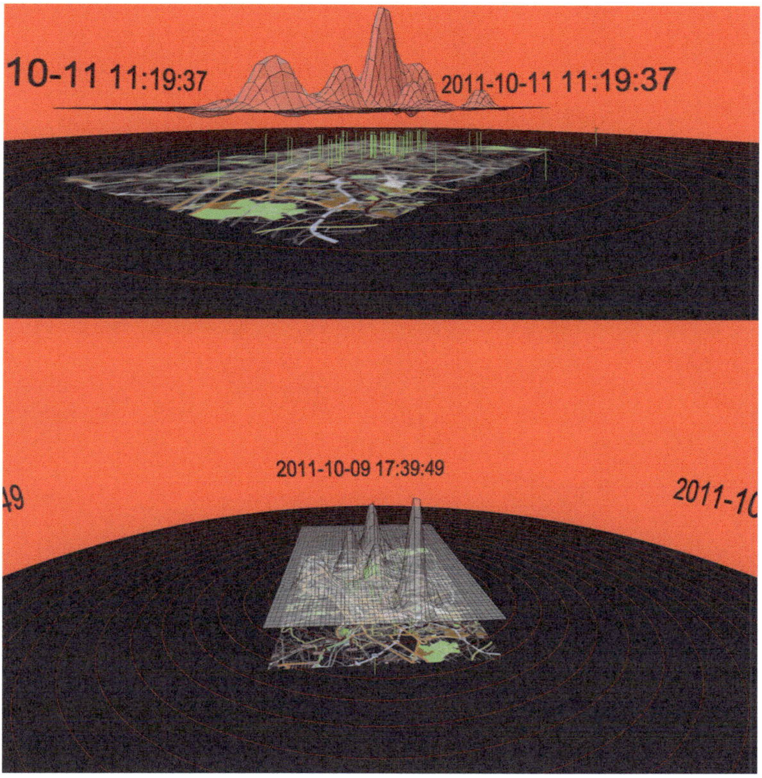

Fig. 5.5 Love in Turin visualized, using a generative video

other. The results were, then, placed onto a printed map, so that the relation with the territory could be appreciated. The idea was to produce such prints periodically, so that a physical museum of the emotions in the city could be produced.

This simple case study constituted our first real-case scenario for the observation of a city through its digital public spaces, and it helped us in the initial effort to formalize what we would call then the Third Infoscape.

One of the results of this case was the realization that one major element was needed in this kind of observation, that is the possibility to observe the relationships manifesting themselves through data: between people, organizations, places, and more. This possibility would enable us to understand the flows of data, information, knowledge, or the propagation of emotions and opinions across the relational links, giving rise to a whole new set of possibilities in understanding the city, and performing it.

This is what the next chapter is about: relations.

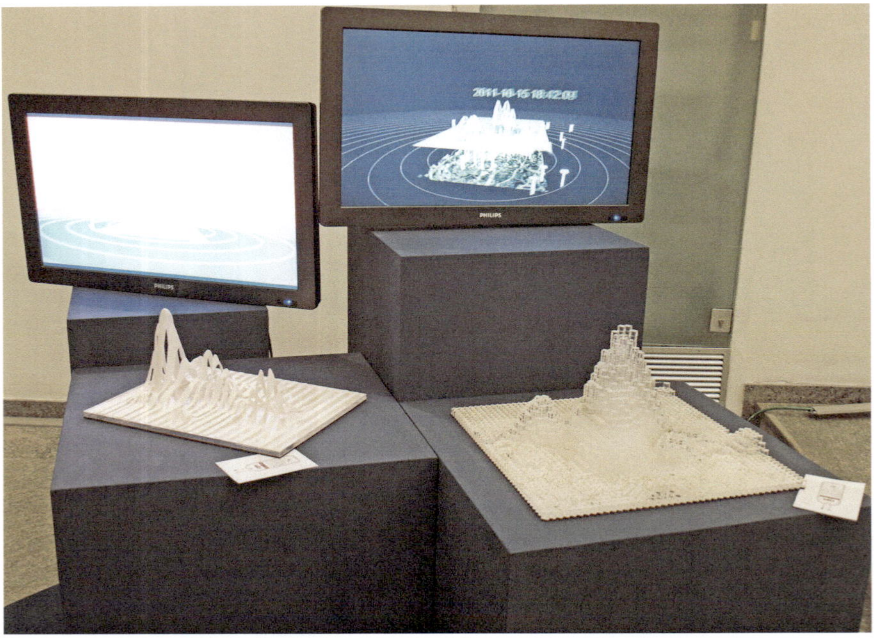

Fig. 5.6 Love in Turin under the form of generative sculptures

Reference

Iaconesi S, Persico O (2015) Visualising emotional landmarks in cities. In: 2014 18th international conference on proceedings of information visualisation (IV), pp 408–413

Chapter 6
The Relational Ecosystem in Cities

Abstract Expanding our understanding of ecosystems: This chapter introduces the concept of the Relational Ecosystem. The urban ecosystems, with all its actors of the multiple human and nonhuman types, seen from the point of view of the relations which emerge and form in its daily life. The chapter introduces the concept and moves on to the description of how relations in the city can be discovered, observed, and measured, starting from the types of data and information which can be captured from the various types of actors in the ecosystems (quantitative, qualitative, classification using ontologies and folksonomies, topic and emotion discovery using natural language analysis, topic/emotion networks and their evolution in time, and relational networks and their evolution in time), and the types of relationships which can be described and observed in this perspective (strength and persistence of relations; vertical, horizontal, transversal relations; liminal relations). The chapter ends with an interview with Massimo Canevacci.

Multiple forms of information and knowledge are exchanged in cities every day, from the most informal to the most formal ones. These exchanges and flows happen across heterogeneous networks which are overlaid onto the city, creating an immaterial geography. This geography is variable across time, location, topics, persistence, and many more variables. Knowledge is the result of the circulation and interpretation of information (Borgatti and Cross 2003), and its creation is a social, interactive, and relational process (Lawson and Lorenz 1999).

These interactions can happen in multiple ways: from asking question to passers-by; to accessing the information and knowledge services which can be offered by companies, organizations, and administrations; to inferring and interpreting information and knowledge directly from the natural and built environment, according to what we see, hear, or feel; and to one's cultures, backgrounds, traditions, and in multiple, other, ways.

Gaining better understanding about the ways in which information and knowledge are exchanged and flow in cities can provide multiple high-valued benefits, ranging from making cities and their opportunities more accessible and usable; to the possibility for easier and more effective collaborative and participatory practices;

S. Iaconesi and O. Persico, *Digital Urban Acupuncture*
DOI 10.1007/978-3-319-43403-2_6

to diminishing divides and augmenting inclusion; to generally increasing the level of readability and usability of cities and of the enormous information and knowledge capital which they host.

Among the fundamental problems of studying the geography of knowledge is that "knowledge flows are invisible; they leave no paper trail by which they may be measured and tracked" (Krugman 1991). This statement is, obviously, partially true, as today's technological innovations have made it somewhat easier to track down the flows of information and knowledge across human networks. Yet it still clearly describes, on the one hand, the ways in which the mere information and knowledge elements and their flows fail to explain how and why knowledge has been exchanged (how it has been interpreted, and if it is possible to provide evidence of people's behavioral transformation when they have obtained it). And, on the other hand, the substantial failure in trying to interpret the more tactical, bottom-up, emergent, and informal ways in which information and knowledge are exchanged.

Here we will try to investigate how digital public spaces and, more specifically, the Third Infoscape can be used to try to assess these two issues, to create a layer of readability of the flows of knowledge and information in the city, to gather information about how, when, where, and in which modality people exchange knowledge and information, what are their roles in this process (in Chap. 8), and how these insights can suggest and promote the emergence of collaborative and participatory practices of multiple types (in Chaps. 9 and 13).

In this process, we will also address the issue of the multiple types of divide (digital, cultural, literacy…) which are found in cities when confronting with technologies.

6.1 Previous Research

We will frame our approach first within general network theory, then by integrating considerations coming from geography and, last, from social network theory.

There is a growing foundation of literature on network theory and knowledge across multiple domains such as economics, organization theory, sociology, and geography (e.g., Borgatti and Halgin 2011; Jackson and Watts 2002; Parkhe et al. 2006; Zaheer et al. 2010).

In particular, Borgatti and Halgin (2011) have provided a concrete approach by creating a twofold model: the flow model, which explains network ties as pipes, leading to capitalization (e.g., the control of information) and contagion (e.g., the spread of information); and the coordination model, explaining network ties as bonds, leading to cooperation (e.g., alliance) and convergence (e.g., co-existence or shared needs).

Within Borgatti's and Halgin's framework, it is possible to characterize network attributes such as positions (e.g., centrality), group memberships (e.g., cliques), roles (e.g., structural equivalence), or dyadic flows between actors (e.g., knowledge exchanges).

Research on knowledge networks has long ignored the geographical dimension (Bell and Zaheer 2007; Whittington et al. 2009).

Recently, scholars have focused on geography as an important factor for organizational change (Freeman and Audia 2006; Owen-Smith and Powell 2004; Sydow et al. 2010).

From the perspective of network theory, geography is a context provider: It creates the environment in which networks exist. Vice versa, looking from the point of view of geography, networks can be observed in their spatial constraints, by asking questions about what geographical characteristics influence the expansion, limiting, enrichment, or impoverishment of networks, their nodes, and links.

To reconcile these two approaches, it is possible to find supporting evidence in those researches which empirically analyze the coincidence of proximity and innovation, for example in Glaeser et al. (1992).

As described in Glückler (2013), four types of models can be found describing the interrelations between networks, space, and knowledge: geography as a condition for network formation; geography as a moderator of network effects on knowledge; network as a moderator of spatial effects on knowledge; network as a mediator of geographical effects on knowledge.

We are going to focus our attention to cities, trying to understand how and if urban environments of different kinds can enable the emergence of networks for information and knowledge exchange, and how and if it is possible to model these exchanges, to gain better understandings about the ways in which they take place and in which they can be facilitated or even suggested.

From this point of view, there are multiple modalities in which research and practice refer to the ways in which technologies and networks in urban settings describe scenarios where information and knowledge exchange takes place.

A first common way refers to the wide range of electronic and digital applications related to the digital spaces of communities and cities (Droege 1997).

Another common reference engages the definition of smart communities, "communities that have made a conscious effort to use information technology to transform life and work within its region in significant and fundamental, rather than incremental, ways" (California Institute for Smart Communities 2001[1]). This perspective has the effect of shifting the focus about city development from "growth through construction" to "growth through interrelation and collaboration."

On top of that, it is also possible to partially merge these two perspectives, when we consider the ways in which sensors, actuators, controllers, and, in general, technologies in the environment become substantially invisible, embedded into physical objects and in the surroundings in which we live, work, express, and move. (Steventon and Wright 2006). This brings on a vision in which the boundaries between digital and physical blur in substantial ways, establishing deep,

---

[1]California Institute for Smart Communities (2001). Ten Steps to Becoming a Smart Community. Retrieved from http://www.smartcommunities.org/library_10steps.htm.

psychological, cognitive links between geographies, human beings, and their relationships which form around communication, information, and knowledge.

One other vision combines technologies, geographies, and human factors in a different way, shifting the focus to the opportunities which emerge when you bring people, innovation systems, and ICT[2] in the same locality—including talented individuals, institutions, and digital innovation spaces—facilitating the emergence of innovation and knowledge creation, sharing, and management (Komninos 2002).

In all of these descriptions, the possibility to construct contexts with high capacity for learning and innovation, depends on infrastructures (from connectivity, to spaces, to urban design elements), relationships, and on the formation of networks, whether physical, digital, hybrid, localized, and global. These are seen as the drivers to leverage the built-in creativity of the population; to create actions dedicated to promote literacy, awareness, information, and knowledge creation; to improve availability, accessibility, and exchange of and with resources, human beings, ideas, and discussions, through institutional, infrastructural, and tactical means.

In this sense innovation is systemic, as described by Lundvall (1992) and Nelson (1993) at national levels, and by Braczyk et al. (1997) and Cooke et al. (1998) at regional and local levels.

Many identify the creation of clusters as the foundation for the emergence of these scenarios (Kaufmann and Todtling 2000).

Different kinds of knowledge flow within these systems, as described by Dawes (2003): declarative (about facts), procedural (dealing with know-how), and conditional (linking conditions and effects); and in its two forms, "explicit knowledge" that is transmittable in formal languages, codified and captured in libraries, archives and databases; and "tacit" knowledge which has a personal dimension that makes it hard to formalize and transmit in other ways than personal communication.

Morgan (2004) has explained how tacit knowledge is spatially sticky, and this quality sustains the trend of innovative activities towards agglomeration.

It is possible, thus, to contextualize these observations within technologically rich environments to understand how, for example, "digital" cities can support the fertile exchanges that are needed for the generation of explicit and tacit knowledge. Ishida (2000) gives a good account of the diversity among the types of "digital cities" which can be imagined and constructed, classifying them into commercial, policy-driven, virtual, and multipurpose digital cities. This classification indicates the scope in which cities become rich in digital technologies: whether it is in support of business and commercial activities in general; or for administrative, bureaucratic, and political reasons, to support education, governance, health, and other topics addressed by public policies; or as in the virtual digital cities, in which cities become characterized by high levels of digital communication and interaction for entertainment, culture, and social interaction, but in which this transformation is not strictly driven by businesses and/or public administrations, but rather emerges through the availability, accessibility, and usability of services themselves, creating

[2]ICT: Information and Communication Technologies.

a different geography in spontaneous and tactical ways, which may or may not be related to the actual physical and political geography of the city. And, of course, these characters of the digital city can be combined, appearing together, as in the case of multipurpose digital cities.

This vision is mostly shared by Schuler (2002), and both describe digital cities as having an information layer, containing data, information, and knowledge; an interface layer, containing the interfaces, maps, and possibilities for interaction; and an interaction layer, where people interact with each other, with institutions, organizations, and companies. This layering allows us to focus on where the transformations and actions take place.

Through these research approaches, we are able to construct a better understanding about how networks-based information and knowledge exchange can work, and how urban contexts can be designed which facilitate (or even enable) the formation of these types of networks. But, we still need to understand how these information and knowledge exchange could actually work, why, using which topographies, architectures, flows, transformations, and modalities.

To gather insights about this process, we can refer to the theories of digital ethnography and of netnography. Both are disciplines which fall at the intersection of ethnography, cultural geography, sociology and more, using different (online and ethnographic) research methods to study the groups, communities, and cultures created through technology mediated social interactions (for example in Kozinets 2006).

These approaches have described models according to which different types of information and knowledge can flow and be shared in communities (or tribes, as defined by netnographers). Examples like Rokka's (2010) research are particularly interesting in our context, as a variant of netnographical methods is proposed to better understand translocal and local phenomena.

Apart from these few initiatives, there still remains an empty gap in the opportunity to understand how information and knowledge are shared and created in cities using social networks, and how these understandings can be used to suggest and promote participatory practices, and the insurgence of a variety of creative and innovative behaviors.

In the next sections and chapters, we will establish a formal hypothesis to be validated against a limited set of research questions.

6.2 The Relational Ecosystem: A Question of Networks, Relationships, and Their Transformations

In the previous sections, we have described the current scenario and the most common understandings about network theory, and its possibility to observe the ways in which information and knowledge are created and shared in formal, informal, and emergent innovation processes, and also highlighted the importance of being able to understand how these networks originate, grow, and transform at local level (especially in cities).

We then observed how the different ways in which researchers and organizations describe the presence of digital infrastructures (material and immaterial, from cables, antennas, sensors, all the way to applications, and social networks) which are able to facilitate or create the emergence of such networks.

On top of that, we understood how the methods of digital ethnography and netnography can help us to gain better understandings about how, where, and why these flows of information and knowledge take place, even linking them to their geographical manifestations, in cities and urban spaces.

With these prerequisites, we can describe the Relational Ecosystem (e.g., a certain city).

Given a set time/space context, the Relational Ecosystem can be defined as a network which evolves over time, and which links the nodes (subjects) within and outside of the ecosystem through the characterization of their relationships, and of their mutation in time.

The Relational Ecosystem is a graph with nodes and links, and it changes over time.

The nodes of the Relational Ecosystem represent the subjects of the network (be them human or nonhuman, such as sensors, databases, organizations…).

The links represent the relationships running between the nodes, their type (e.g., describing whether one node feeds information to another one), persistence, strength, and other characteristics (which we will see next).

The Relational Ecosystem has boundaries which we set through an initial hypothesis, by choosing what/who/where/when we want to observe: a city and its citizens and tourists; a neighborhood; the discussions happening on social networks over a certain topic; a specific number of people and organizations. These are just some of the examples.

The boundaries defining what is in or out of the Relational Ecosystem are dynamic and fuzzy. While it is possible, in general, through the initial hypothesis to understand what/who is in or out of the ecosystem, it may soon become evident that something which is initially "out" should really be brought "in," as the levels and degrees of its relationships with the rest of the ecosystem could justify such inclusion. This may happen frequently on social networks.

For example, consider the scenario presented in Fig. 6.1: the observation of the cultural life of the city of Rome (as we will see in Chap. 10). In our observation, we could include cinemas, to understand what online discussions, expressions, and conversations influence who goes to see the movies in which cinemas. Through observation, we construct a model in which thousands of people, in Rome, talking about movies and cinemas, compose the Relational Ecosystem for this topic, in the city. All of the nodes in the ecosystem will not only have conversations and interactions within the ecosystem. For example, a number of them may follow a movie review blog (and its Facebook page) to know more about upcoming movie releases. The person behind the blog may be from anywhere, maybe from Colombia. We may put these interactions (e.g., judging from how users in the

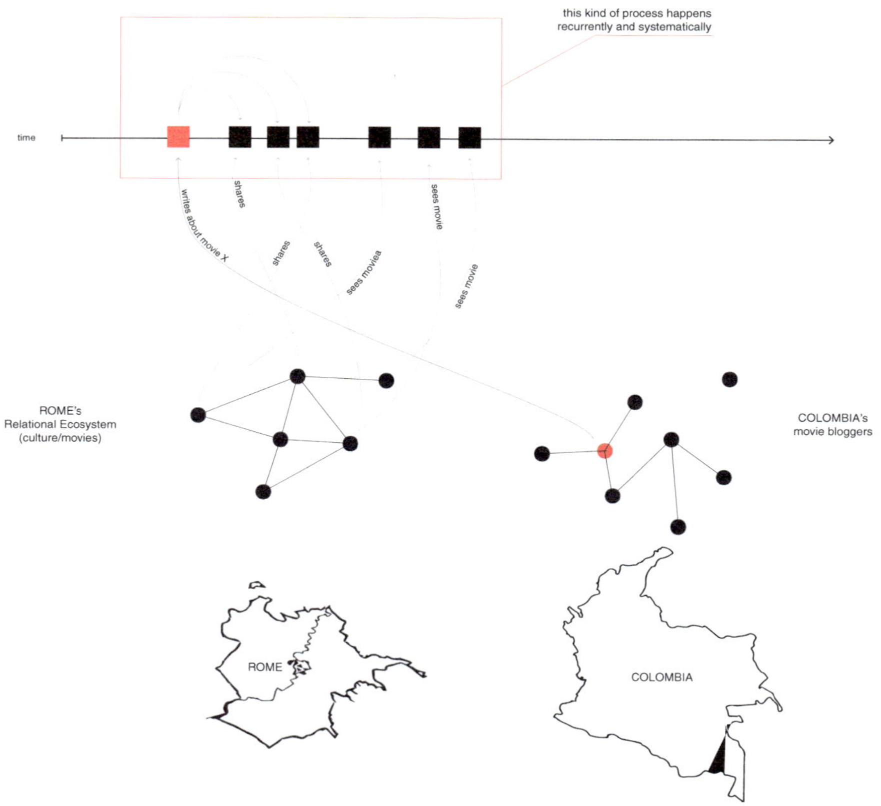

Fig. 6.1 A fictional example scenario to understand the boundaries of Relational Ecosystems: cinemas in Rome

ecosystem share information from that blog) on a timeline, and then observe how people express that they have gone to see certain movies. Now, if the blogger systematically influences how people go to see movies in Rome, should the person/blog/page be inside or outside of the Relational Ecosystem in the city of Rome? Obviously there is no single answer, as it depends on what we want to achieve. We could highlight this systematic, effective influence as an external factor (to evaluate, e.g., how the city's cultural life is directed and influenced from external agencies), or we could include the node (the person or the blog) in the Relational Ecosystem, and maybe visualize it in a map, to show how well interconnected the city of Rome is in the international discussions about cinema. Any of these and more options would be feasible and interesting, for different purposes. If the blogger from Colombia was effectively and systematically able to influence Rome's movie marketplace, he could be considered as one of Rome's cultural operators, to all effects, even if he had never been to Rome even if he does not have a company or organization in the city. This new geography would reach out and stretch all the way to Colombia, and the relations would go with it.

With understandings and boundaries set, is it possible to define a classification of the types of connections and nodes of the human and nonhuman Relational Ecosystem in urban spaces with sufficient presence of social media activity? One which is able to highlight in meaningful ways the processes according to which data, information, and knowledge are created, shared, exchanged, and manipulated in the city?

And, if it is possible, how can this type of classification, and its manifestation during the everyday life of the city, be used by citizens, administrations, organizations, companies, and by artists, designers, and creative classes to facilitate the emergence of designs, new services, participative, and collaborative initiatives, peer-to-peer actions, civic actions, and participatory and collaborative decision-making and policy-shaping processes?

In the following sections, we describe a methodology that was used to attempt at providing answers to these questions.

Of course the types of answers to these kinds of questions may vary according to the context and to the scope of what we want to achieve. Nonetheless, it is for sure possible to observe the digital part of reality; to try to understand how this part of reality relates to the parts which do not find expression in the digital realm; to learn useful patterns which may come of use in physical, digital, and hybrid domains; and to use these types of knowledge, information, and data in inclusive ways, so that people and organizations can use them directly or as inspiration to make things happen in the city by collaboration and participation.

6.3 Characterization of Data and Relationships

To perform these tasks, we have to set forth some goals, characterize the data which is collected, establish a methodology to collect and process the data, and to be used to gather the insights and understandings which we are looking for.

The focus is, again, on emergent information and knowledge in the city, on the ways in which their flows describe the life in the city and on how it is possible to use this knowledge to create participatory, collaborative interventions.

As we have seen from Ishida's and Schuler's research, it may be of interest to understand at what level the various processes take place at the information layer, the interface layer, or the interaction layer. This is of fundamental importance for a variety of reasons: from the visibility of the data, information, knowledge, and of their flows; to the capacity to understand who controls the layer and, thus, can influence its politics.

As we have seen from Borgatti's research, it is useful to comprehend the models of network processes: the flow model, leading to information capitalization and contagion; the coordination model, leading to cooperation and convergence.

In this, it is useful to classify the types of flows which can take place, for example using Dawes' research: declarative; procedural; and conditional. Dawes can also be used to differentiate explicit knowledge from tacit one, and as it is determined through network processes.

In the end we can consider how geography comes into this, for example by using Glückler's classification: geography as a condition for network formation; geography as a moderator of network effects on knowledge; network as a moderator of spatial effects on knowledge; network as a mediator of geographical effects on knowledge.

Thus, a possible scheme for characterization of relations emerges, shown in Table 6.1.

The phenomena which is the object of this analysis is the emergence of relations in the city (or "around" the city, as we have seen in discussing the definition of boundaries for the Relational Ecosystem), as they become manifest through digital technologies of multiple types (social networks, sensor networks, transaction systems, communication systems and services, etc.).

Thus, we will be dealing with complex systems of dyadic relations. Each relation is a dyad: two nodes are engaged, with some form of relation, and characterized according to the parameters of our interest (if we are interested in movies, as in the example in the previous section, one parameter might be how many movies have the two subjects seen together). These bundles of dyadic social relations come together and form a network. Individuals, whether they realize it or not, are entangled in thick webs of relations which manifest themselves through our jobs, schools, purchases, energy consumptions, romantic relationships, families, entertainment activities, and more.

Table 6.1 Characterization of relations

At what level does it happen?	Layers	Information
		Interface
		Interaction
How does it happen?	Flows model	Capitalization
		Contagion
	Coordination model	Cooperation
		Convergence
What kinds of things happen?	Types of flows	Declarative
		Procedural
		Conditional
	Types of knowledge	Explicit
		Tacit
Where does it happen?	Geography	Condition for network
		Moderator of network effects
	Network	Moderator of spatial effects
		Mediator of geographic effects

The material we will be working with is the network: a concrete set of nodes (each one corresponding to a person, an organization, a sensor, a system, a service, or another form of data-subject, a subject which is able to emit or absorb data); and a concrete set of ties, links, which connect nodes, two at a time.

Nodes can exist at different levels.

Some of these levels are hierarchical, meaning that some node can contain several different nodes: imagine an organization, which could contain business units, each containing the people who work in them; each level could express on its own or in terms of its members (e.g., the company could have its own presence on social networks; a certain business unit, maybe a prestigious laboratory, could have a presence on its own; while the members of the laboratory could each have their own profiles on social networks, whose expressions may not necessarily be connected to the ones of their unit or company).

Others are more fluid, meaning that the characteristics which one node must has to be considered part of the node of the superior layer can be more volatile, ephemeral, dynamic, and ever changing: for example, if I was to participate on a social network discussion about rock music, I might be considered part of the "rock community" super-node while I am participating actively to the discussion; the moment I stop (maybe because I have even deleted the whole discussion from my records, for all that the others know) I will not be part of that upper level node.

As we have mentioned, nodes can be "human" or "nonhuman" (e.g., a sensor, or an organization, or one of the more recent smart-refrigerators, who may establish a relationship with me on social networks by reminding me the shopping list of urgent supplies I would need to purchase). We will observe some of the implications of this condition in Chap. 11.

Within all of these types of relations, there are formal and informal aspects. Formal aspects regard intended, expected, and also "authorized" or "compulsory" relations: they are the ones typical of organizations and bureaucracies, or even with services, in which there is the customer, the provider, the procedure, the manager, etc. Formal relations usually involve roles which are static, or which change slowly through time and space. Informal aspects involve unexpected, volatile, trust-based, respect-based, humor- and irony-based interactions, such as gossiping, helping, informing, assisting, and joking (which, of course, also happen within the strictest corporations, not only in the streets; and, as internal communication managers for these organizations know, they have just as much impact of the life of the companies as the formal interactions, if not more).

What comes out is a non-reductionistic, holistic, sui generis, emergent-properties analysis of social phenomena, which is very powerful and effective: not really focusing on what elements are in the system, but on how well they are put together.

At this point, we need to characterize the data we collect and the information we are able to extract from it from all of these perspectives, to be able to determine a set of variables which will allow us to classify the relationships in the Relational Ecosystem. A possible scheme for applying these classifications is shown in Fig. 6.2.

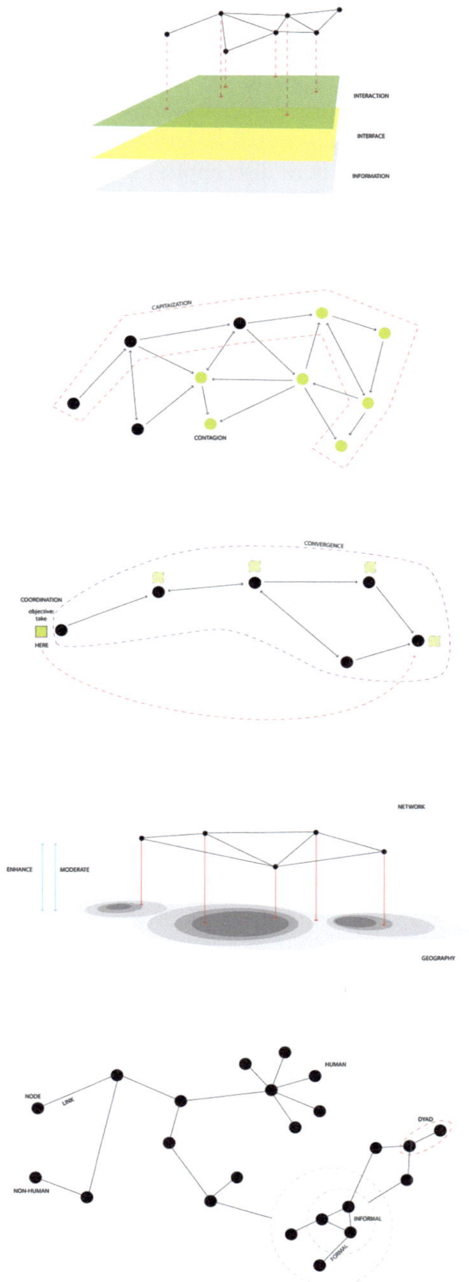

Fig. 6.2 Ways in which to apply the classification of relations in the Relational Ecosystem

All of these—and more—relation characterization processes will have to be performed based on the data that is captured from the Third Infoscape of the city.

When it comes to capturing data from the environment (human, natural, and social), the current scenario offers many issues to be confronted with. Let us start with describing possible harvesting processes on social networks, knowing that what we say here also applies to other, common, situations, in which a few large data managers (service providers, big data companies, organizations, and more) share control of the (quasi) totality of the available data, forcing us to face similar issues.

Harvesting large amounts of public messages and content from major social media services (such as Facebook, Twitter, Instagram, and Foursquare) poses both legal and technical issues.

Starting from the legal issues, users and developers wishing to use the features of major social networks have to abide to the rules dictated in the providers' Terms of Service (ToS). These are very complex legal documents which state what can be done and what is prohibited, also establishing various forms of liability for all parts involved.

Different providers have different ToS, which can vary substantially in describing the ways in which you can and cannot use the information generated on their services.

In the specifics, most of our focus was oriented towards the ToS documents offered for developers.

Most social networks offer Application Programming Interfaces (API) of some sort, which developers can use to build their own applications by interacting with the social network's ecosystem (users, communities, content, data…). These APIs offer an incredible opportunity for service designers and developers, as they permit accessing a vast amount of data about people's expressions and positions, the topics they discuss and the relations which they maintain, allowing for the creation of a variety of useful services.

APIs usage is constrained by the ToS, which limits the degree to which any developer or company is able to capture, process, use, and visualize information coming from social network operators.

Limits are mainly imposed on ownership of the data, number of queries that can be executed over a period of time and on the numerosity of results that can be provided, storage and processing of the information, and possibility to freely use the information.

These legal limits are different across different providers and also change quite frequently and arbitrarily, forcing companies and researchers to constantly adapt and maintain their applications: if your application is perfectly ToS compliant and working today, it might not be so tomorrow.

Limitations can be found on the technical side as well.

In the first instance, the APIs allow for limited degrees of freedom in the querying and interaction with the databases of operators: not all of the information is made available and limitations on how developers are able to formulate the queries also exist.

Furthermore, APIs frequently change, forcing development teams to constantly maintain and adapt the source code of the applications.

Once in a while, entire sets of features and possibilities disappear or change in form or availability, forcing designers and developers to go back to the drawing board and rethink or reframe their services.

It can be said that the ideas of accessibility and of interoperability are currently not among the priorities of social networking service providers.

In our selected solution, we have chosen to access publicly available APIs whenever possible/feasible, and to refer to established data providers (such as Gnip, or directly to the service providers) whenever, larger amounts of data were needed.

These approaches, for example, are implemented in the Human Ecosystems technologies. To overcome many of these issues we have also adopted the Ubiquitous Commons scheme, which is allowing us to not only weather down existing legal and technical problems, but also to reflect on the critical issues which they imply, creating positive impacts on people's rights and their possibility to express and to organize themselves autonomously.

The harvested data is processed in multiple ways to achieve different results. In this phase, we used the approach described in (Iaconesi and Persico 2014), to obtain the geographic localization of content (through GPS enabled smartphones, geographic named entities extraction, or other techniques), to understand the topics and emotions discussed and expressed in the harvested content, and a characterization of the relations established through them (e.g., inferrable from the fact that they shared, forwarded, expressed appreciation on some other content, or through the other multiple types of actions which are made available on the different social networks).

We use the data extracted from public profiles and content to aggregate:

- quantified data;
- quantifiable data and categories;
- topics/emotions networks; and
- relational networks.

The relations between these types of data are shown in Fig. 6.3.

Through social networks, it is possible to collect a variety of quantified data: how many times people interact with each other; how many friends/followers/relations they have; from which geographical location they write; and how many times they discuss a certain topic or express a certain emotion, according to which time sequences and more.

This type of data is often used to weight other data/information, or to perform quantitative analyses of certain phenomena, or even to guess the relevance and reliability of certain other information (e.g., if a certain news item or information is constantly mentioned with a certain emotional expression by members of a community which is relevant to the topic, this could be a measure of how relevant is that news item for the community, much more than a single person mentioning it just once).

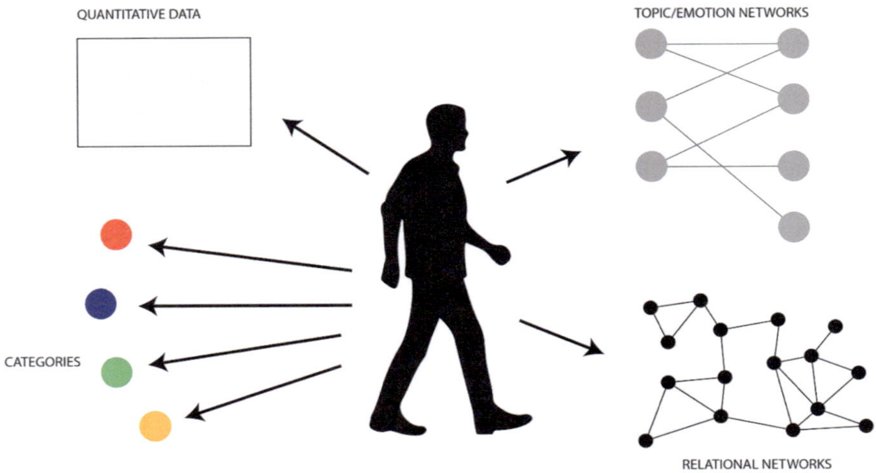

Fig. 6.3 The schematic representation of the types of data collected from social networks

Quantifiable data and categories enable to categorize items, in more (ontologies) or less (folksonomies) strict ways.

For example, if someone would mention on a social network that he/she graduated from University with an Engineering degree, we could add them to the category of people with a university degree and tag them with "engineering" in our folksonomies. If, some times later, they would mention in a public message on social networks their interests for, let's say, robotics, we could tag them with "robotics," and so on, collecting information as we move on.

These first two types of data and information can be seen as a peculiar form of census. Peculiar in the sense that they are different from the data which can commonly found in, for example, national census, for a number of ways, for example:

- they are updated in near-real time (which positive, but not always good: for example, it is not uncommon to find data which has been generated the previous day, compared to data generated the previous year; these types of issues can be solved by extensive usage of time-stamping techniques);
- they are not certified (it is not uncommon that people do not tell "the truth" on social networks; or, better, people tell multiple version of truth and reality on social networks; this is dangerous for census, and it is, instead, highly desirable for anthropological, cultural and perception analysis, because it describes people in the way in which they choose to represent themselves, their desires, fetishes, and more);
- they are not intentional; this last fact has implication both for data quality and usability, and for issues like privacy and surveillance; from the first point of view, information generated non-intentionally is seldom relevant, and often not focused on the exact "questions" we are looking answers for; from the second

point of view we refer back to the previous discussion on the Ubiquitous Commons, by noting that a limited amount of people is actually aware that the information they generate may be used for multiple, unexpected purposes (if I publish a picture of holidays at the beach in Bahamas, this information may render me a target for marketing campaigns); for both approaches we follow strict methodological and ethical rules (e.g., employing anonymization, strict quality assessments), together with the possibility for novel visions (e.g., the focus on "emergence" rather than the analysis of strictly characterized phenomena; of by including in our research wide, accessible, and engaging education and awareness processes, through which we educate citizens to protect themselves, to learn how to participate actively, in precise ways, and how to use this data and information themselves).

Further analysis is dedicated to how individuals discuss multiple topics and express their emotions about them.

The topics which people discuss are captured as well as the ways in which they relate to each other. For example, if a person talks about Topic A and Topic B, it could point out that these topics are related in some ways, or that people talking about Topic A could be also interested in Topic B. Of course, these possibilities must be properly evaluated, and multiple types of assessments of this kind are possible: the fact that this type of information is available opens up the possibility for these kinds of analyses to be performed: by analyzing the topics of the discussion we can hypothesize a number of ways in which to suggest collaboration, participation, practices for mutual support, civic communication, and more, understanding relations between citizens which citizens might not even know exist (e.g., citizen A cares about a certain topic but does not know that citizen B does too).

The same can be said about emotions. Emotions are detected using the natural language analysis techniques,[3] in relation to the topics about which they are expressed, in their sequence in time (also opening up the possibility to understand how emotions evolve over time), in their connection to news and events (e.g., trying to understand if a certain news item might be connected to a transformation of the emotion about a certain topic for a person, or a community, as shown in Fig. 6.4).

These, combined, form complex networks, which contributes to better understanding about how information, news, knowledge, emotions, and opinions are created and flow across the city through people, communities, and cultures, for example, by simple clustering algorithms acting on time, space, emotions, and topics.

In these few paragraphs, we have already seen multiple ways in which relations among data subjects can emerge. This gives rise to the possibility of understanding

[3]For a detailed technical and technological description of the techniques used we suggest to look up our "Urban Acupuncture in the Era of Ubiquitous Media.", which contains very detailed descriptions of the algorithms, software tools and instrumentation which we use for this process. You can look it up in the references at the end of the book, or grab it on our website at http://www.human-ecosystems.com.

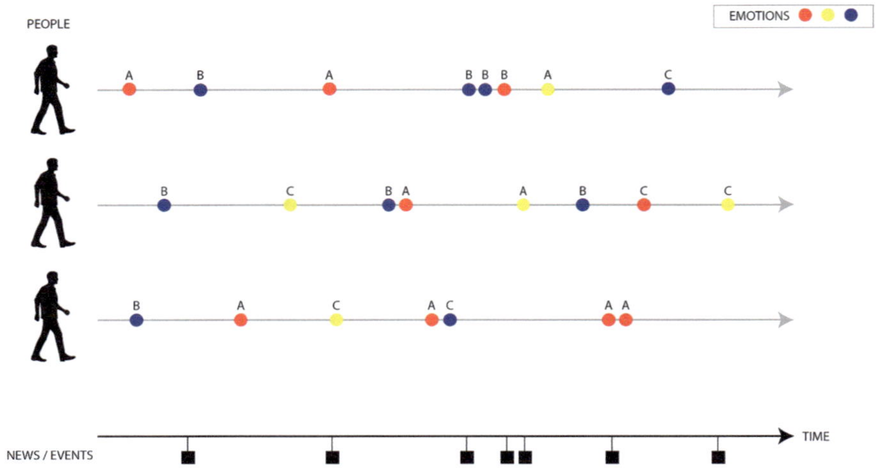

Fig. 6.4 Studying the possible correlations between topics, emotions and news items

more about people's relational networks, observing how they can originate and evolve, in several different ways, and according to multiple modalities.

They can form by direct or indirect relation, or around discussions on specific topics, news items, knowledge elements, information bits, expertise, and more.

A relation can form by retweeting a content, commenting on a post, sharing some information, mentioning someone, and talking about a topic of interest. All of these modalities contribute to the formation of each person's and organization's relational networks, describing the ways in which they are interconnected, and in which they evolve in time.

Not all relations are the same. Possible parameters which can be used in discerning the differences among relations are shown in Fig. 6.5.

For our purposes, we have taken into consideration the following characterization of relationships:

- weight: the strength of a relation; a retweet is weaker than a conversation, for example, with each interaction adding up to form the weight of the relationship;
- persistence: the duration of a relation; one casual message exchanged among two people is very different from a continuous state of conversation, across time, domains, and interests; persistence allows to understand how reliable the relationship is, how it extends in time; how constant and present it is;
- verticality: some relations are vertical, meaning that they are focused on a single topic/domain (or a limited set); for example, they are the ones found in focus groups, or forums; this measure is an aptitude, an evaluation indicating how much the relation is prone to sticking to a certain topic; it is initially less meaningful and gains more profound significance with the augmentation of the relation's weight and persistence;

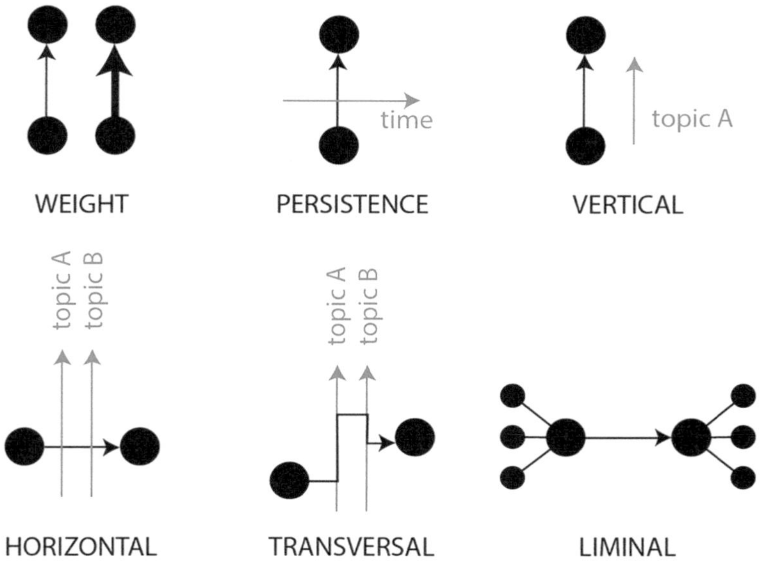

Fig. 6.5 Some of the parameters which can be used to distinguish different types of relations

- horizontality: some relations are horizontal, meaning that they are not focused on a certain topic/domain, and they engage the same people across multiple topics; this is the case of friendships, for example, in which the people engage and possibly discuss a very diverse set of topics;
- transversality: these are relations in which both horizontal and vertical characteristics are manifested; sometimes they are vertical, sometimes they are horizontal; sometimes the two modalities originate one from the other (whenever conversations start around a certain topic and end up other ones, in ways which do not permit deciding whether the relation is horizontal or vertical); this is the case which is used in social networks include different communities in a certain analysis, by finding convergence on a series of topics, and by researching the divergences and differences; this is a measure which becomes relevant for persistent relations, with the possibility to analyze them over time, as conversations shift from one topic to another, or embrace different topics at the same time; and
- liminality: liminal relations happen on the border of something; they are relations which can connect or separate; they describe borders and transition spaces, interstices; they are specifically interesting for their characteristic of being across something, in-between, defining the ways in which something is connected or separated from something else, for example describing the bridges or separations across different communities or cultures; liminality is described in a series of different ways and is measured through an aptitude, a ratio measuring how liminal the relation is; the possibility to measure how much a certain node

interacts with two different clusters of relations with high persistence is possibly liminality's most useful effect, because it allows to discover bridges between different communities.

All of these characterizations have a lot to say with respect to the types of networks which we have described in the beginning of the chapter.

They will be able to describe on which layer the relation forms. People discussing something will relate to the interaction layer. A relation born through a feature of the user interface (e.g., if a certain user was to comment on a "related post" which appears on some social network interface) they would be relating on the interface layer. Two users relating algorithmically (for example because some algorithm discovered their affinity through their similar reactions to a set of news items on a certain topic), they would be relating on the information layer. In this last case, the two users might not even know that they are related.

They will be able to describe the relations according to the flow and coordination models: for example, to the ways in which some users are able to capitalize flows of information by controlling multiple relationships or by the ways in which they are able to achieve contagion of a certain opinion or emotion, through the relations which they can activate or they could describe how people achieve their goals through collaboration (maybe getting introduced to someone or accessing an opportunity) and convergence (maybe by reaching consensus on a certain issue across a whole subnetwork).

They will also be able to describe the types of flows and knowledge which manifest themselves across the network. Simple discourse analysis, structurally understanding the forms and architectures of the texts exchanged, may reveal whether discussions are declarative, procedural, or conditional: for example, if user A poses a Yes/No question, user B answers it, and user A does something as an effect, it would be a conditional flow, easily identifiable through the analysis of the structure of the text. Or, in regard to the classification among explicit/tacit knowledge, a declarative flow may characterize explicit knowledge, while a network which expresses itself on a certain topic without anyone having started a declarative flow may be one of the signals of the existence of a tacit knowledge ("everybody knows, but no one says").

On top of that, spatial considerations may provide further insights about these processes. For example, if the nodes are geographically close, we will be able to observe their interactions and flows by analyzing how their closeness becomes an enabler for more persistent and ponderous networks, or how they resist to network effects coming from spatially distant locations.

Each of these characters constitutes a value which can be measured and analyzed to search for patterns, aggregations, and tendencies, which enable us to characterize the Relational Ecosystem which we are observing.

In the next chapter we will see a case study, in which this approach was applied to the observation of a city and its innovation ecosystem. And, then, we will move forward to the possibility to analyze some recurring patterns of these relations, to attempt a definition of some roles for the network's participants.

6.4 An Interview with Massimo Canevacci

Massimo Canevacci *is an Italian ethnographer and critical thinker. He is well known for his works about contemporary metropolis, "native" cultures and digital arts. Professor Canevacci is currently based on the Institute of Advanced Studies of the University of São Paulo, Brazil, and he has been living in this city since 2008. He is married to the artist Sheila Ribeiro. Before moving to Brazil, Canevacci was a Professor of Cultural Anthropology and the Faculty of Communication Sciences of the "La Sapienza" University of Rome. He is particularly well known for the development of the concepts of "communicational metropolis," "multividual" as well as the tensions between "self-and hetero-representation". He is also known for his contribution to the study of digital cultures and his dialogic relationship with "native" populations such as the Bororo and the Xavante of Central Brazil.*

Ubiquity. When digital cultures arise, fragmented spaces, times, identities, perceptions, representations, and contexts can interconnect and recombine, spontaneously, consciously or even unconsciously, without realizing it, for example, through algorithms and computation. This is a radical transformation. In your opinion, what are the major cognitive, social and political impacts of this possibility?

Typically, different human and social sciences expressed a sense of horror towards everything which seemed to be fragmented, not homogeneous, spontaneous, which affected the concept of identity—always seen as fixed, certain, compact, and homogeneous—and the paradigms based on the variations of dialectics. From another point of view, the digital revolution was considered only as a matter of technology: these theories did not comprehend its profound cultural, communicational, and political implications.

The process which started disruptively by the middle of the 1990s manifested through the practices of diffused subjects an immaterial geopolitical scenario which was truly different from the one which was radicated in hegemonic systems. On the other hand, artists, writers, anthropologists, and architects had announced and affirmed the perturbative diffusion of the simultaneous recombination of the fragments which were dispersed in the metropolitan flows, and which were interweaved with the gaze of the artist. For example, through collage and cut up, but also through the method of collaborative montage by different authors, independently one from the other: Thomas Mann, Walter Benjamin, Gregory Bateson were inspired by classical music, metropolitan experiences, by epistemological narratives.

The composition of the musical languages influenced Mann who, in an exchange with Adorno, the secret counselor of Doctor Faustus, said in 1945:

"What I want particularly to account, commenting myself, is the method of installation, running through (…) the whole book."

So Mann, in the composition of his literary work, applies the leitmotif invented by Wagner—a sort of acoustic assembly pasted into any character—in his writing short sentences that introduce the entry of each character. A method which Mann claims as philosophical rather than literary.

A few years before Walter Benjamin, author loved by the writer of Luebeck and whom unfortunately he never met, working on his unfinished work or maybe impossible to accomplish in the Bibliotheque National in Paris for his research into the Passagen and writes in a fragment of almost random text:

“Method of this work: literary montage.”

Around the mid-30s, Benjamin is working on the issue of mechanical reproduction, both of Paris, capital of the nineteenth century metropolis experiencing communication as a feature of modernity. The experience of the assembly is lived in the urban streams of the nineteenth century, where photography (and then film) was born, not casually, with Daguerre, to become the exhibition method which was suitable for his research. Filmic montage, urban montage and literary montage intertwine in a unique capacity for exhibiting through fragments.

Before and after these two authors, Gregory Bateson—who represents theoretical perspectives and disciplines which are completely different—leads ethnographic research in New Guinea with the Iatmul where he learns the limits of the anthropological exhibition of rituals. The text—Naven—is still considered as experimental, as the author immediately understood that the explanation of a cultural trait, in this case a ritual for sexual inversion, brings crisis to the epistemological systems. In the postface from 1958 he writes:

“Explanation is about the fitting together of data.”

The accusation of “selecting facts to fit my theories,” at the time was diffused and Bateson tried to avoid it by writing in an experimental manner. For this, since the beginning, he stated that the true method is not so much about what happens in the fieldwork, but the one which takes place when going home and writing. In the following years, Bateson develops the concept of metalogue, in which a conversation must make relevant not only the interventions of the participants, but also the structure (pattern) of the entire debate. The metalogue is a montage of spoken word. Metalogue is “fitting together of data” in the form of discourse. A montage of sentences and facts.

Finally, the Anglo-Iraqi architect Zaha Adid said in an interview about diagonal forms as the beginning of her design: “I always think about how we can put the object together.” And, further on, she notes that her works “may well have anticipated the computer revolution that has allowed my office to now design very complex structures. I think the design method can be considered as the precursor of computing.” So through her composition of the digital world “we see a proliferation of natural morphologies and entities that might be imagined as a living creature.”

To conclude, Mark Weiser is an author who has anticipated the emergence of digital ubiquity as one of the great problems of today: the potential of ubiquity, the flexibility of identities and the interconnections which go beyond the object–subject dualism extend the nineteenth century subconscious to the pixel-skin of the PC-screen, or of the other mediums. This is the fundamental passage.

Another impact of digital cultures is the establishment of the concept of Composition as one of the driving principles of communication. Remix

cultures, generativity, computation, but also other human processes such as fashion, entertainment, relations, sexuality and politics are a matter of performative composition, of a continuous state of research and experimentation on the ways in which "things" can be combined, connected, interweaved, intertwined. This gives rise of a conception of relationality which is completely new.

How can the concept of Composition drive the discovery of novel forms of relationality among human beings?

This is where my proposal for the Digital Montage Methodology (DMM) presents itself within the cultural syncretisms in the form of a transitive constellation performed through the "fitting together of data" and by "putting the object together." Whereas it is clear that data and objects incorporate fragments of subjectivity. The constellation is a travelling sideral concept which is composed by different planets, satellites, meteorites which wander in a space which is in movement, not fixed. In its diversified movements the wandering constellation includes visual configurations, compositive narratives, empirical contexts, irregular logics, incongruous crossings. It is a syncretic constellation in which fragmented concepts form a montage during the research and tune into a polyphonic composition which is not based on a written musical score with the notes inscribed onto the pentagram. The composition of the syncretic constellation follows the indications of contemporary music, in which the score is a sort of musical design which every executor develops according to her own aesthetic sensibility. In this sense, the syncretic constellation and the polyphonic composition draw acoustic movements which are dissonant and which try to express visions which are physiognomic of the compositive syncretism, made by polyphonies, etheronomies and ubiquities.

Mirò is able to express the sense of the constellation. Mirò is the astonishment of art. His pictorial and conceptual works achieve the possibility to present a composition of fantastic expressions—true and unreal—which move the observer and affect him towards mutations which are self-compositional. In the constellation Mirò paints the imagination of an enigmatic eros, indisciplined, transitive. The result is the floating dance of wandering beings, like identities and methods. Mirò's Constellation draws the musical score which is fit for polyphonic composition. Music in the form of painting. Painted essay. Performative ethnography. The astonishment of the composition-constellation which moves towards possible syncretism.

Identity is a concept to be handled with care. Not only it can only find definitions through differences, through the continuous, fluid mutation of the networks of relations among people, environments, objects, memories, projections, expectations and representations, but it also needs to take into consideration the beyond-human consideration that identities are also defined in unexpected, unintended, temporary, recombinant ways: by companies, algorithms, administrations, events, contexts, artificial intelligences, digital transactions and more. Not only we are multividuals, but also recombinant, temporary, anonymous, collective, connective, nomadic, transitory, according to the ways in which we are constantly remixed and recombined (inside,

across, transversally) for the purposes of business, control, entertainment, sociality, politics, sex, work and more.

What is your point of view on the emerging possibilities for "identity"? What do you think are the possible futures of identity in this scenario? How can we use these insights to create a more inclusive, joyful, relational world?

A smart question already contains possible answers. For example, the concept of identity has been analysed by anthropologists (like Paul Rabinow and George Marcus) and has been the most used concept during the past decades in doctorate thesis all over the world without bringing any innovative vision. The two authors think—and I agree—to lead to a decay of the concept, to imagine new ones. Here again artists can be useful. I propose the concept of hetheronym and of eus developed by Ferdinando Pessoa. His poetry interferes with the labyrinths of the solidity of the I, which is so impermeable and single-minded in its most notable manifestations, to melt the rigidity and to liberate its potential multiplicities. Hetheronym practiced in the various forms of writing is not only a literary device, and not only a psychological one. Hetheronym assumes different identities and unveils through the action of self-naming narrative modules which are diversifiable, never groupable in the unified synthesis of the public-private subject, and which are fluent in the many paths which precariously depart, traverse and assemble the nexus of subjectivity-style-perception-vision-listening-elaboration-composition.

Hetheronym is the polyphony of the I which overflows in the different writings.

Pessoa incorporates polyphony and lives it in the radical choice of embodying the being-author which challenges his writing corporality through polyphony, the subjective anxiety of not being constrained in one's own monolithic individuality and, thus, becoming a multividual: that which he defines as os outros eus.

His other Is are the writing corporalities of the poet. The solved node of the I. This expresses the perspective of os outros eus. The solved nodes of the Is. These anticipations, in my opinion, interweave (digitally syncretize) with the tendencies of things, which are never any more only objects.

Ubiquity determines contemporary subjectivity in relation to digital communication. During these last few years this term has been applied to identify a modus operandi which uses web cultures and digital design and which has gone forward in this connection: the web is ubiquitous and the communicational ubiquity characterizes the space-time relations of the Internet. Ubiquity follows a sensorial logic with a material/immaterial character; it expresses immanent tensions beyond dualism that simplified feeling of the human condition in which binary oppositions are functional to reconducting the complexity of the ordinary into the dicothomic domain of the instrumental.

Ubiquitous is the potential of fantasy which connects with technology. The subject of the ethnographic experience is ubiquitous and multividual.

The concept of multividual (beyond the indivisibility of the individual) is fully manifested within these ubiquitous interconnections. Ubiquitous expresses tension beyond the fixed identifications of things and beings, to offer enlightened poetic-political visions. For the ubiquitous subject the space-time coordinates become unnecessary and superfluous, or, better, transitive. This experience does not mean the s-materialization of the interpersonal relationships: it presents optical, manual, mindful and imaginary connections which move even in the apparent immobility of the subject. The concept of connective multividual manifests itself in these ubiquitous connections through the fragments of space-time which are void of predetermined identifications or normalities, and which multiply the experiences based on temporary identities.

To be able to use (and benefit) from all of these concepts, new types of imagination and new forms of conflict are required. New forms of imagination, to embrace the form of continuous, ubiquitous performance which is needed to recognize new forms of beauty (found, for example, also on dissonance, and in difference) which can be used construct the scenarios of the future, polyphonically, using multitudes of points of view. New forms of conflict, to establish novel dynamics, not based on contrast and contraposition, but on composition, ecosystems and difference.

In your experience and research, what are possible, viable and sustainable new forms of imagination and conflict? What is the role of artists, designers, architects, poets in their definition/enactment/performance?

This is the most painful side of current politics which, in the residual and immortal movements of the twentieth century, perpetuates a model of conflict which is obsolete as the ghosts of the great taylorist industry. Industrial and conflictual archeology. Incapacity to experiment the decisive relation between the forms of conflict and the mutant material/immaterial context. Some holograms to protest against the government in Madrid go in this direction. But here we need an Exact Imagination to radically innovate the nexus of conflict-innovation.

To be able to experience and experiment all of these concepts, we must be able to observe and experience them, or, in other words, to develop a sensibility for them. This fact points in the direction of a type of research which cannot take place inside laboratories, but which must happen in a state of continuous and participatory performance. On top of that, this kind of research should be diffused, including the ones which would be, traditionally, "observed" into the performance of the experiment and of the research. For this, we derive the term "P2P Ethnography" (Peer-to-Peer Ethnography) from Ethnography: a diffused participatory observation in which myriads of subjects are able to capture and represent microhistories, transform them into a commons, and perform them through art, education, citizen engagement, and tactical uses.

What do you think of this term/definition and, more in general, what do you think the future (or present) of research should/could be, with specific focus on the possibility to understand the flows of data, information, knowledge and communication to gain better understandings about the ever changing Relational Ecosystem of cities?

The concept of performance has extraordinary compositive, multividual and conflictual potential. In performance the design of the structural difference between actor and spectator becomes archeological, an aesthetic division of work which reproduces an ossified domain. Digital tendencies clearly go in the directions in which each subject-thing is a performer, active participant to innovation and interconnection. Spect-actor is maybe an ugly term and it does not offer the potentialities which are prefigured and in which the connections among the diverse forms of art-design-architecture-music etc. graft into diffused activations which are multiplied in urban ecologies. Here the concept of ecology is the one from Bateson reenacted in the digital, recalling how he was invited by Wiener to elaborate on the concept of feedback or schismogenesis in the body of cybernetics. Mental ecology which includes nature in its human and technological mutations and also in its animal ones, herbal, and rocky. An applied relational ecology which is decentered and performed in the material/immaterial spaces of the metropolis, which is my project for an ubiquitous, non-anthropocentric anthropology.

References

Bell GG, Zaheer A (2007) Geography, networks, and knowledge flow. Organ Sci 18(6):955–972

Borgatti SP (2011) On network theory. Organ Sci 22:1168–1181

Borgatti SP, Cross R (2003) A relational view of information seeking and learning in social networks. Manage Sci 49:432–445

Braczyk HJ, Cooke P, Heidenreich M (1997) Regional innovation systems. UCL Press, London

Cooke P, Morgan K (1998) The associational economy. Oxford University Press, Oxford

Dawes PL (2003) A model of the effects of technical consultants on organizational learning in high-technology purchase situations. J High Technol Manage Res 2003(1):1–20

Droege P (1997) Intelligent environments—spatial aspect of the information revolution. Elsevier, Oxford

Freeman JH, Audia PG (2006) Community ecology and the sociology of organizations. Ann Rev Sociol 32:145–169

Glaeser EL, Kallal HD, Scheinkman JA, Shleifer A (1992) Growth in cities. J Polit Econ 100:1126–1152

Glückler J (2013) Knowledge, networks and space: connectivity and the problem of non-interactive learning. SPACES Online, 11(2013-02)

Iaconesi S, Persico O (2014) Urban acupuncture in the era of ubiquitous media. J Community Inf 10(3)

Ishida T, Isbister K (eds) (2000) Understanding digital cities. In: Digital cities: experiences, technologies and future perspectives. Springer, Heidelberg

Jackson MO, Watts A (2002) The evolution of social and economic networks. J Econ Theory 106:265–295

Kaufmann A, Tödtling F (2000) Systems of innovation in traditional industrial regions: the case of Styria in a comparative perspective. Reg Stud 34(1):29–40

Komninos N (2002) Intelligent cities: innovation, knowledge systems and digital spaces. Spon Press, London

Kozinets RV, Belk RW (2006) Netnography 2.0. In: Handbook of qualitative research methods in marketing. Edward Elgar Publishing, Northampton

Krugman P (1991) Geography and trade. MIT Press, Cambridge

Lawson C, Lorenz E (1999) Collective learning, tacit knowledge and regional innovative capacity. Reg Stud 33:305–317

Lundvall BA (1992) National systems of innovation. Towards a Theory of Innovation and Interactive Learning. Francis Pinter, London

Morgan K (2004) The exaggerated death of geography: learning, proximity and territorial innovation systems. J Econ Geogr 4(1):3–21
Nelson RR (1993) National innovation systems. A comparative analysis. Oxford University Press, Oxford
Owen-Smith J, Powell WW (2004) Knowledge networks as channels and conduits: the effects of spillovers in the Boston biotechnology community. Organ Sci 15:5–21
Parkhe A, Wasserman S, Ralston DA (2006) New frontiers in network theory development. Acad Manage Rev 31:560–568
Rokka J (2010) Netnographic inquiry and new translocal sites of the social. Int J Consum Stud 34 (4):381–387
Schuler D, Tanabe M, van den Besselaar P, Ishida T (eds) (2002) Digital cities and digital citizens. In: Digital cities II: computational and sociological approaches. Springer, Heidelberg, pp 71–85
Steventon A, Wright S (2006) Intelligent spaces: the application of pervasive ICT. Springer, London
Sydow J, Lerch F, Staber U (2010) Planning for path dependence? The case of a network in the Berlin-Brandenburg optics cluster. Econ Geogr 86:173–195
Whittington KB, Owen-Smith J, Powell WW (2009) Networks, propinquity, and innovation in knowledge-intensive industries. Adm Sci Q 54:90–122
Zaheer A, Gözübüyük R, Milanov H (2010) It’s the connections: the network perspective in interorganizational research. Acad Manag Perspect 24(1):62

Chapter 7
A Case Study: Kansas City's innovation ecosystem

Abstract Greater Kansas City's innovation ecosystem, as observed in 2012–2013. The chapter explores Kansas City's innovation ecosystem, as observed for two full years through major social networks. It begins by explaining what motivated the authors in choosing to observe Kansas City (its lively, young and growing innovation scene, its startup and technological hubs, and the organizations working on the territory) and then moves on to describe the methodologies and techniques used in the observation, and to visualize and highlight some of the results: the quantitative and qualitative aspects, over time; the transformation of the innovation ecosystem over time; the communities, hubs, topics, emotions, and drivers of the ecosystem; and the accessibility and inclusiveness of the innovation ecosystem and its impacts, measured throughout communication, social impacts, and the emergence of patterns and cycles.

The dataset for the Greater Kansas City's case study is available at http://www.human-ecosystems.com/DAP/DigitalUrbanAcupuncture-CaseStudy-KansasCity-dataset.zip

During our Eisenhower Fellowship[1] in 2012, while in Kansas City, we visited the Ewing Marion Kauffman Foundation,[2] to discuss innovation, education, and about the possibility for creation of effective innovation ecosystems: environments in which innovation and entrepreneurship could prosper through education programs, networking activities, grants and support for young entrepreneurs, and more.

Kansas City is a peculiar place for innovation. It is far from the beaten tracks in this domain: for example, it is distant from the west and east coasts' hubs for technologies and startups, both perceptively and in practice. But decades of investments, together the welcoming environment of the city (which is sufficiently small, not too small, and also significantly cheaper in terms of cost of living in respect to Silicon Valley, New York, or other innovation hubs), have transformed

[1]The Eisenhower Fellowship is an international program which was born as a birthday gift to president Dwight D. Eisenhower, to promote a "more prosperous, just and peaceful world." To know more about the Eisenhower Fellowship: https://www.efworld.org.
[2]Ewing Marion Kauffman Foundation: http://www.kauffman.org/.

S. Iaconesi and O. Persico, *Digital Urban Acupuncture*
DOI 10.1007/978-3-319-43403-2_7

the place into an effective ecosystem for innovators, startup entrepreneurs, and technological companies development. Entities like the Kauffman Foundation and KCNext[3] are right at the center of this process.

During our conversations, we started wondering if it was at all possible to observe the communication processes and the relational activities of all of the subjects involved in Greater Kansas City's innovation ecosystem to gather insights about its well-being, its effectiveness, its aptitude in establishing nationwide and international networks, and in creating impact. We were also wondering if it was possible to grab these insights and use them to design meaningful interventions in the ecosystem, to improve its performance.

We decided to use this scenario as a test case for our ongoing research, to explore practical application of our methods.

We decided to start from trying to find ways to observe the following types of expressions:

1. Kansas City as an attractor of innovators, to try to evaluate how the city captured the attention of entrepreneurs, investors, organizations, and researchers as a place for innovation;
2. Kansas City's internal relational dynamics, to assess how collaborative, and cooperative dynamics evolved in the area;
3. Kansas City's external relational dynamics, to explore how the subjects within the city were able to relate to people in other places, at national and international levels, to disseminate their findings, to establish partnerships and networks, to find markets, and to create impacts.

We used different data harvesting modalities for each of these three items. The observations started at the end of 2012 and ended on September 30, 2013. In the following sections, the indication "social networks" indicates Facebook, Twitter, Instagram, Flickr, and Foursquare.

Let us examine them.

7.1 Capturing Data

(a) Kansas City as an attractor

For this part of the observation, data were captured from social networks by searching for mentions about Kansas City (and about a limited number of influential organizations, spaces, and individuals in the area), generated by subjects who are not local to Kansas City, and which also mentioned interest in the area in ways which are compatible with the "attractor" pattern, on which we were investigating.

Let us analyze these elements.

---

[3]KCNext is an affiliate of the Kansas City Area Development Council dedicated to making Kansas City a leading destination for technology companies and professionals.

Mentions about Kansas City and its influential subjects was performed by simple full-text search: we would capture whenever people mentioned the city, or locations in its area, some of its organizations, institutions, companies, influential subjects or events, and set it aside for further analysis. To do this, we combined simple full-text searches, based on dictionaries (of places, names, organizations and more in Kansas City), and checking, through software-based discourse analysis algorithms whether the mentions were compatible with their referring to Kansas City (e.g., using the way sentences were built to understand whether the person speaking had been to KC, or was interested in it, or was going to be there, through grammatical, structural, and morphological analysis[4]).

The next step was purposed to understand the locality of subjects and their tendency toward Greater Kansas City.

For all of the users identified in the previous step, an estimation of their "home location" was performed. This took into account a variety of parameters:

- the user's expressed "home location," as it is declared on social networking profiles (where available); this was deemed as important, but not too important, as people often state imprecise or even false or fictional locations (e.g., "Mordor," coming from "The Lord of the Rings," which we found frequently);
- the user's geographical distribution of posts; if at least 50 % of the user's geo-referenced posts are from a certain place and the others come from multiple locations, this fact would highly contribute to the determination of the user's "home location."

In this phase, users whose "home location" was not in Kansas City were selected.

The third step was performed to identify user as innovators, entrepreneurs, investors, leaders, policy-makers, researchers, hackers, makers, designers, and creators. This was performed by analyzing the typical content which they posted. For all of the users which passed the previous filtering step, their social timelines were inspected to try to find content which we imagined could be relevant to the types of subjects we were looking for: technology discussions or articles; innovation, business, startups, disruptive business models, and more. To do this, we performed natural language analysis on the texts they posted, and analyzed the links that they shared, against a number of relevant websites and domains.

The fourth step was to identify users as interested in the city in relevance to its characterization as an "attractor." For this, we identified tendencies toward Greater Kansas City: "coming to"; "investing into"; "moving to"; "keeping an eye on"; "know more about"; and similar phrasings, denoting a tendency and attention toward the area.

The result of this progressive filtering can be seen in Fig. 7.1.

[4]As in all of the other parts of the book: it is always possible to look in the GitHub repository of the Human Ecosystems project on https://github.com/xdxdVSxdxd/HumanEcosystems to see (and use) the actual software tools which have been used to perform this and the other experiments.

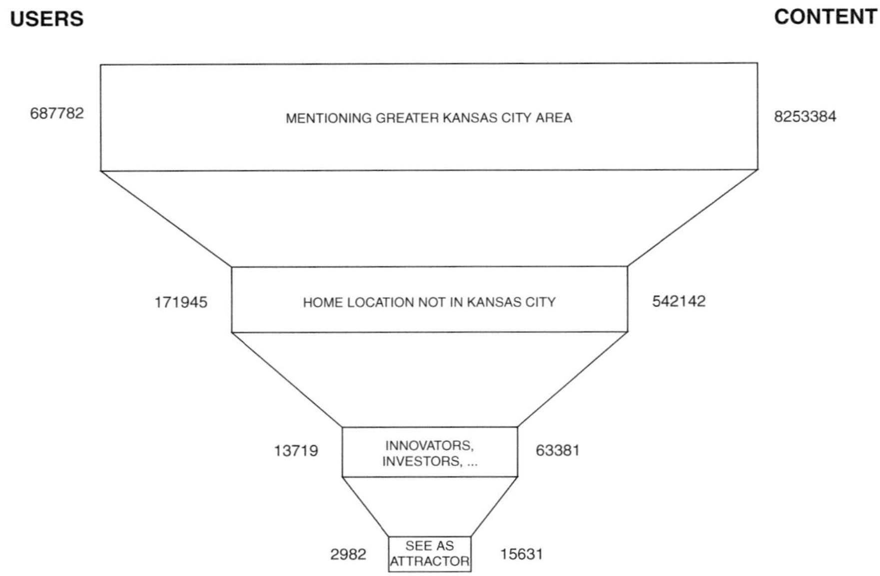

Fig. 7.1 The number of content and users captured in the "Kansas City as an attractor" modality

Figure 7.2 allows us to appreciate where these users were located in the world.

(b) Kansas City's internal relational dynamics

The objective of this part of the observation was to learn how the city's innovation ecosystem communicates among its participants, to understand its flows of communication and interaction, to model the ways in which collaboration and knowledge exchange take place, and to try to infer which interventions could be prefigured to improve its performance.

To achieve this, the observation was configured in the way explained in the next few paragraphs.

Initially, the users posting on social networks from the Greater Kansas City area about innovations, startups, research, social innovation, industry, finance and markets, and a few other topics were detected. This was done by simple word and hashtag matching in posts harvested within the relevant geographic area.

These detected users were put under observation, by capturing their updates. Of these, the ones which were relevant for the topics of interest were maintained. These, in turn, were also monitored, to see if they generated any interactions with other users. The ones which were encountered within the Kansas City area were added to the observation, to the set of monitored users. The others, reacting from other geographical areas, were added to the observation of the external relations, as seen in the next section.

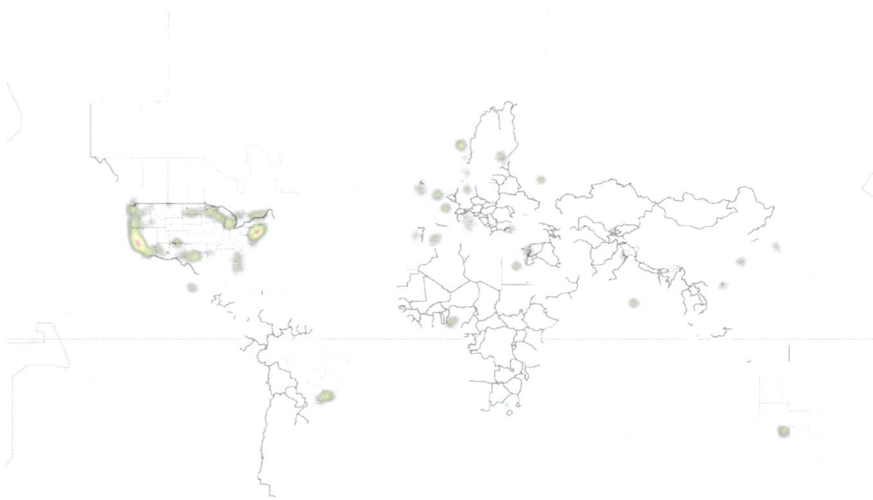

Fig. 7.2 The locations in which users perceive Kansas City as an attractor for innovation

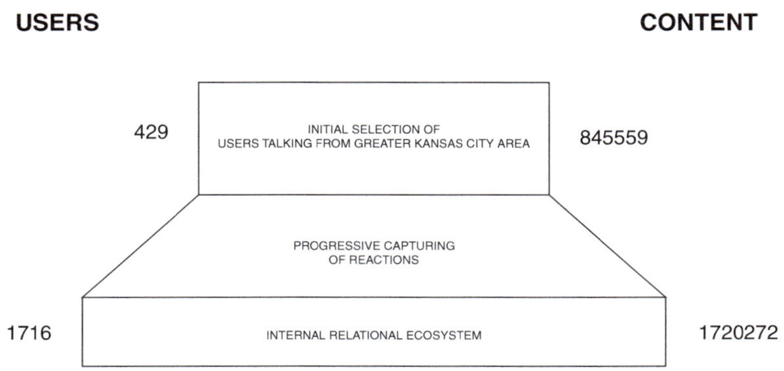

Fig. 7.3 The number of users and content captured for the internal Relational Ecosystem

This became an iterative process (harvest, check relations, add to monitored user base, and repeat), progressively expanding and fine-tuning the observation. The results can be seen in Fig. 7.3.

Relations within this user base were detected either directly or indirectly.

Directly, through retweets, shares, comments, and other forms of direct and explicit social interaction. Each different type of interaction was valued differently, as to measure its contribution in term of relation weight, as seen in the previous section (e.g., a retweet is weaker than a comment, but if a certain user retweets

Fig. 7.4 The relational graph of the internal Relational Ecosystem in Kansas City

systematically all of the tweets of another user, this may mean that a strong relationship exists, even under the form of a partnership; these and other considerations were applied).

Relations were discovered indirectly through other mechanisms, for example, by monitoring which users shared the same links, or discussed about the same news items, or expressed similar emotional responses to the same stimuli. These, as well, were attributed different weights.

In this way, the nodes (users) started to become connected through links having different weights. With their evolution in time, it was possible to also make considerations about their persistence vertically/horizontally/transversally/liminally. The internal Relational Ecosystem can be seen in Fig. 7.4.

In the end, a final evaluation was made (in the last section of this chapter).

(c) Kansas City's external relational dynamics

The last point of the investigation was similar, in modality, to the previous one. The major difference falls in the logic according to which users were discarded. If in the previous section we saw only users writing from the Greater Kansas City area maintained in the observation, here the opposite is true: the ones publishing their expressions from other areas (nationally and internationally) were maintained and monitored. The filtering process can be appreciated in Fig. 7.5.

The graph of the Relational Ecosystem was laid out in a similar way, but highlighting the internal and external relations, so that they could be immediately and visually inspected.[5]

---

[5]These kinds of graph are incredibly hard to read through an image. They need interaction and motion to be interpreted correctly: filtering, digging, and zooming become fundamental actions, to

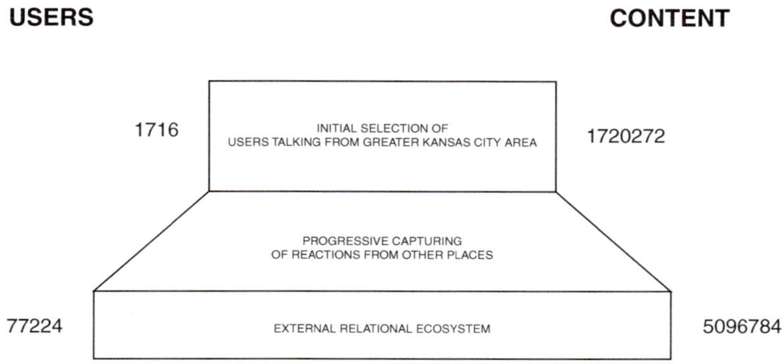

Fig. 7.5 The number of users and content for the external Relational Ecosystem in Kansas City

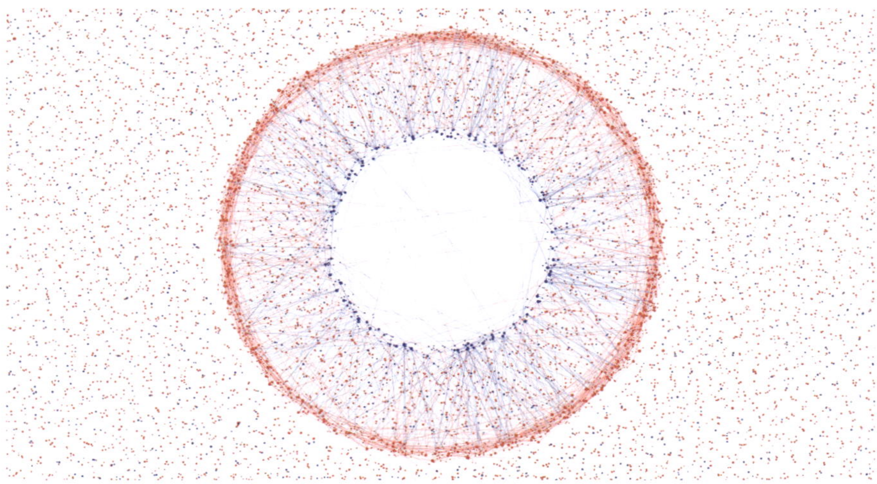

Fig. 7.6 The 8000 most persistent relations in the relational graph of the external Relational Ecosystem in Kansas City

The relational graph and the geographic distribution can be seen in Figs. 7.6 and 7.7, respectively.

(Footnote 5 continued)

be able to inspect these graphs at the internet address http://www.human-ecosystems.com/DAP/DigitalUrbanAcupuncture-CaseStudy-KansasCity-graph-external.zip you can download the software code used to produce this visualization. It is written in the Processing programming language (www.processing.org). It connects directly to the dataset, and includes zoom and dig functions, and the possibility to save the searches to an animated video, like this one: http://www.human-ecosystems.com/DAP/DigitalUrbanAcupuncture-CaseStudy-KansasCity-graph-external-video.mov.

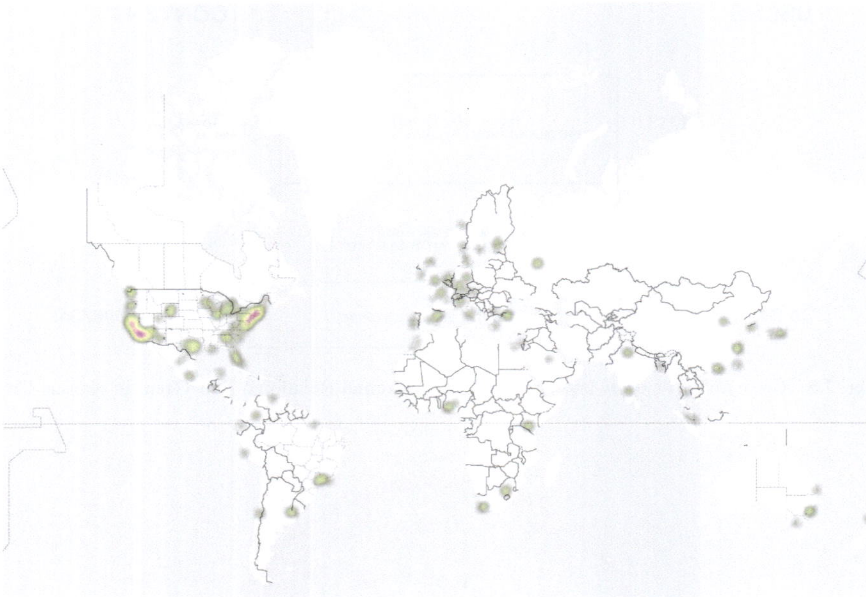

Fig. 7.7 The relational geography of the external Relational Ecosystem in Kansas City: places which have relations with subjects in the Kansas City area

7.2 Evaluation

What can we understand from these datasets and visualizations?

Let us start from the first set: Kansas City as an Attractor.

Some insights can be gathered right from the numerosity of the harvested content. Of almost seven hundred thousand users mentioning the Greater Kansas City area, about 170 thousand are from outside the area itself. That is a little more than 1 in 6. That is not too bad, but it is, still, a small number: it means that of 7 people mentioning KC, only one is from outside KC. The area talks a lot about itself.

1 in 12 of these people not in Kansas City are innovators, entrepreneurs, investors, researchers, and start-uppers. About 1 in 6 of these have expressed, at least once, that they perceive KC as an attractor for innovation, that "there's something there," that it is worthwhile checking out and, maybe, going there to see what is happening with innovative practices, technologies, research, and business. Adding those up, about 1 in 450 people who mentions the Greater Kansas City area online, mentions it in ways which allow perceiving it as an attractor for innovative practices.

How could we intervene in this figure? In one of many ways.

First by intervening on who does not yet mention Kansas City, by giving them something to talk about and appreciate, something relevant, and relevant for innovation. We could, for example, do the opposite process: we could start looking at people discussing innovation, and "contact them" in some way (even a simple Twitter mention) to propose them some news, issue, or discussion in which to take part.

Then, we could intervene on the ratios of innovators versus non-innovators who talk about Kansas City. We could get more people from innovation communities interested, by providing them something relevant to discuss.

And then we could intervene by taking the innovators who do not speak about KC in terms of an attractor for innovation, see what exactly they are expressing, address it, and let them know that we have addressed it. Or, even better, we could even ask them to help us in addressing it, to get them engaged.

As we saw in the previous sections, we have also produced a map, showing where in the world people seeing KC as an attractor for innovations are. The resulting map is reasonable enough, allowing us to finding many of the things we would expect: people in Silicon Valley and New York; people in Brazil; people in Northern Europe and Spain; people in Nigeria, with their exponentially growing digital economy; India, China, and Japan.

What could we do to make the situation better? A number of things. We could translate things, and communicate them in different languages, referring to specific people. We could engage relevant people from the countries we are missing and that we did wish were there. We could talk about their initiatives, comparing them to the ones started in Kansas City, and bringing up discussions.

Let us move on to the internal dynamics of the Relational Ecosystem.

As before, the numbers provide us with some initial insights. A little more than 1700 people discussing innovation, in one of many ways. That is almost an army. They produce more than one and one half million messages each year. That is more than one thousand messages each; almost 3 a day, every day, Sundays and holidays included.

How are these people using these conversations in regard to promoting collaboration, knowledge exchange, mutual support, and discussion in the area?

A simple visual inspection of the graph visualization leads us to a quick consideration: many people (the nodes, the dots in the graph) are alone, isolated. They may be "speaking" with someone else, maybe outside KC. But they are many. If they show up as isolated in the internal graph in the Relational Ecosystem, it means that they never once mentioned someone from the area, commented on one of their posts, shared the same link, forwarded their content to their communities. Maybe they did it passing through some accounts whose expressions on social networks are protected. But they still are many.

The possibility is that they do not know each other; or that they know each other, but they think they have nothing to say to each other, even if they deal with innovative "things"; or that they just do not care. Or many other possible explanations.

Why should we want to change this situation? One reason is that more collaboration and communication could represent a tremendous competitive advantage. Imagine if your Twit had 1716 retweets, or if each post you shared was forwarded by that many people to their community, or if each time you had a problem you could count on almost two thousand potential helping hands with incredible expertise, or if you could access all of these innovators for coming up with an idea, brainstorming, testing your latest App, validating some of your research, or finding a market for it. And we could go on in listing the advantages.

What could we do to change this situation? We could look, for example, at what people are talking about, especially the people which show up as isolated dots on the graph, and connect them, maybe through a simple mention on social networks, or inviting them for a shared meeting or event. We could provide tools, shared hashtags, Facebook pages, and other means through which collaboration could become easier: a tool for sharing within the ecosystem, layered on social networks; a known set of hashtags or Twitter lists with which to publish something, advertise your new service, ask for help, find a researcher or a developer, search for partners, and more. And we could communicate them: after all, those dots are people, and we can easily contact them using social networks.

The same kind of considerations goes for the external dynamics of the Relational Ecosystem.

If we look at its graph visualization, we will see many isolated dots (the visualization on paper shows only the 8000 most persistent relations in the external ring you might want to use the digital tool or watch the video, as indicated in the footnote, to see the complete set). Again, we could set up tools, methods, hashtags, pages, and meetings or events in which to learn how to leverage this wealth of relationships, to advertise, communicate, find help or partners, discover interesting research, and create markets. We could communicate these tools, methods, and practices in physical meetings or events, or online, leveraging social networks, so that it would become easier for people in KC to access the networks of all of the other people in KC, constituting a fundamental competitive advantage.

To do all of this, we would need to gather some deeper understandings about the roles which different subjects have in the Relational Ecosystem, so that we may think how to make better use of the people who create connections, the ones who influence opinions, the ones who amplify the reach of messages, the ones who are able to put people in contact with each other, the ones who curate all the information smog and select relevant messages, and so on.

We would need to understand people's roles in the Relational Ecosystem, and the resulting typical flows of information and knowledge. We will do it in the next chapter.

Chapter 8
Understanding the Relations and the Flows of Emotions, Information, and Knowledge

Abstract Understanding the roles in the Relational Ecosystem, this chapter introduces a methodology according to which the nodes of the Relational Ecosystem can be classified according to their characteristics, and the ways in which it is possible to define further classification schemes to achieve different goals. The types of relations described in Chap. 6 are used to understand the ways in which they can be analyzed to classify the members of the ecosystem according to multiple schemes, to be able to provide meaningful insights about the ways in which data, information, knowledge, emotions, and opinions flow in the city: information brokers, hubs, experts, influencers, amplifiers, bridges, and more. This chapter describes how these and more roles can be identified for human and non-human members of the ecosystems (organizations, e.g., or objects, plants, environments connected through sensors, or even data and information sources of various kinds which may be present in the ecosystem).

In the previous two chapters, we have looked at how information about the unfolding and mutating relations can be captured to describe the Relational Ecosystem of the city? What data, information, knowledge, emotion, and opinion do the various subjects of the ecosystem emit and exchange? What flows take place? What can we understand from these flows? How can we design interventions to promote the well-being of the ecosystem? How can we design the overall system for it to be inclusive, not only related to the people who use social networks, smartphone, domotics, sensors, wearable technologies, and other devices or technologies?

While exploring this domain, we might have noted four facts:

1. This field of exploration is constantly moving across the boundaries which separate ethical and unethical behavior, and law and regulations do not help;
2. In the Relational Ecosystem, not all participants are human beings;
3. In the Relational Ecosystem, geography is a "strange" concept;
4. To really understand what goes on in these complex scenarios, we need to be able to attribute roles to subjects, in computational ways; these roles may vary

S. Iaconesi and O. Persico, *Digital Urban Acupuncture*
DOI 10.1007/978-3-319-43403-2_8

according to time and context, but to be able to understand what happens, we really need to understand who/what is contributing to causing the flow and contagion of information, emotion, and opinion, who is amplifying, who is connecting, and so on; roles.

8.1 Ethics and Civic Action

About fact number one.

Systematically capturing data and information about how people relate to one another and about the ways in which they interact with technologies, services, and organizations may seem a scary thing to do, potentially violating privacy, rights, and liberties. And it is.

The problem is that these kinds of procedures happen millions of times each minute, through the constant action of organizations and their algorithms, that this fact is not clear, perceived, and transparent, and that not only it is impossible for users of these services to know what is happening and how, but also that it is impossible for citizens to express their will about it and to actually use all of this information for their own purposes.

In regard to the ethical issues involved in these types of harvesting processes, we adopt (and suggest adopting) the following approach:

- **Only public:** We act on multiple levels to make sure that the information which we capture and generate is derived only from those messages, images, data, and information which users have explicitly published as "public," meaning that the fact that they published it in the way they did is a sufficient guarantee that the "message" is for everyone; if I publish some image on Facebook as "public" ("everyone can see"), for example, this image is shared in a way for which it gets indexed from search engines, and everyone can access it, even if they are not logged into Facebook (e.g., it can be shared on other social networks, or through its link). At another level, if I were to publish a certain message in which I would lament the poor condition of the potholes in the streets in my neighborhood, my objective would not to keep such information private, but to make it public, to share it with my fellow neighbors, with the street maintenance service, with the mayor, and with other citizens. On yet another modality, if I configure my mobile application to say that I would like to share the information which it generates for civic or scientific uses (maybe it is a mobility application, whose information can be used to gather information about traffic; or an health application, connected to a biomedical device, through which I can publish information which is useful for scientific research), that also would be an explicit statement declaring the fact that I wish that my information be made public. These and other cases are the ones whose models we use to try to infer whether messages and information are "public" or not. Security and privacy configuration patterns, message comprehension to understand whether it is intended for the public or not, and modalities in which to harvest for messages, these and others are only some of the ways in which we try to make sure that the data and information which we are capturing is meant for the public sphere or not;

- **Open**: Make the process and the data as open as possible. We make the algorithms and software products which we use completely open source (in fact, they are distributed under a GPL licensing scheme), so that people can actually know what we are looking for, how we are capturing it, how we process it, and how we classify things. So open, in fact, that people would be able to actually use them for their own purposes, if they wished, or even to sue us if they thought we were doing something wrong. Even the data are published in this way. As described when we introduced the Human Ecosystems project, the data captured and processed from the city become a source of Open Data which can be accessed from anyone. In these two ways, we can address one of the main issues in this area: opacity. Have you ever seen Facebook's algorithms? Or the algorithms and the actual perspectives on your data which they use to categorize you as "heavy metal fan," or "loves cats," or "homosexual," or "terrorist"? This opacity has dangerous implications, which we address through openness.
- **Enable**: Even all of this care dedicated to capturing only public content, and making the process open and transparent is not enough. The "wrong" information will eventually arrive on systems like these, even due to issues which may be out of our control. Someone will make mistakes in their privacy settings. Someone else will publish some information about someone else without obtaining their consent. Someone will not realize that by publishing information in the way they did, it is as if they published it onto the New York Times. And let us not forget the fact that most of the information we generate, we generate without realizing it, unconsciously. Or, on top of that, someone may wish to avoid having their information used for any of these purposes, because they do not want to. The possible cases are endless. For these reasons (and, also, to adhere to laws and regulations, e.g., in the European Union), we consider a best ethical practice to offer citizens clear, understandable, accessible, usable indications to understand what information is actually and effectively available about them on these kinds of systems, to download and browse/use it, and to be able to remove it, if they wish to do so. When we say remove it, we actually mean to completely remove it, in all its aspects and direct/indirect contributions, including the information about the request for removal, and the possibility for data from the same subject entering the databases again.[1]

[1]This may seem as a paradox: How do I prevent data about a certain subject to enter the database, if I don't store any information about that subject? How can I avoid data from or about Mr. X entering the database if I don't store Mr. X's name somewhere. There are techniques to do that in ethical ways. One possibility is to store Mr. X's name in ways which are not accessible to us, so that we cannot know whose subject data we are filtering out. One way to do this is to act when Mr. X requests deletion of his/her data: When this happens, a cryptographic key is generated, used to store the information about Mr. X which we need to filter out his/her content from future

- **Educate**: Teach people how to make sense of all this, and how to use it and control it. As we saw when we introduced the Ubiquitous Commons and the Real-Time Museum in the Human Ecosystems project, we dedicate specific effort to provide people with legal/technological tools and education processes to be able to take active control of this scenario, and also to learn how to use this scenario, individually and, more importantly, together with their community, with the rest of the citizens, with their neighbors and friends, with the people that care for the same things, to promote the emergence of a better, more aware, active, collaborative type of city. Of course, in fact, it is not enough showing people just how much data they generate: because most of them would have no clue in understanding what to do with this situation. They would only become angry, or frozen which fear and anxiety, which is possibly even worse. This is because most of the people do not have the culture, vision, imagination, and skills to understand that (1) they are actually the only ones who do not have any access to the data which they generate as a collectivity and (2) that these data can be used for positive, constructive, and productive ends. Through inclusive education programs, and through the wide diffusion of tools such as the Ubiquitous Commons, it becomes possible to build these cultures, visions, imagination, and skills which can completely transform the scenario, by transforming citizens' imagination of the city and, thus, their desires, expectations, and competences. Citizens do not only need to understand how to protect themselves, but also need to understand (and learn) how to use all of this wealth of information, learning how to construct, collaborate, and communicate with it. This is one of the most crucial aspects of the approach described in this book.
- **Participate**: Do not act like you are in a sterile laboratory and take action with people. This kind of research should never be done in a laboratory. It should be done with people, together with citizens, as part of participatory actions in which decisions about what has to be done, or has to be researched, or observed should be decided together with the communities, groups, cultures, and people who are being observed. "Let's observe ourselves, to learn more about us, and to understand how to make our lives better." It is definitely a different message than: "We have observed you, and understood what you do and how you behave in your daily lives." The first one can be joyful and liberating: It is a self-observation of an engaged community wishing to learn more about itself, to understand how to change for the better. The second one is a statement on the news, like when they raise your taxes: if you were not engaged in the decision, if the decision comes from above, and if there has been not even a try to get you involved in finding better options, in making you understand the problem and

(Footnote 1 continued)

harvesting processes, providing Mr. X with this key (so that, eventually and if he/she wants, they can use it to reverse the process, or to check whether everything worked out as expected) and to store it in a way which is unusable to us, and only accessible by the software (e.g., using hashing techniques with randomised keys, which only the software knows; another, more secure, way, is through the Ubiquitous Commons protocol, described at http://www.ubiquitouscommons.org/).

participate in finding the answer, you won't like it. This is, also, another crucial element of the methodology described in this book. Which is a methodology about caring and about doing things together with the people who live in cities, not about observing cities like an enlightened voyeur with a digital, infographic dashboard.

All these modalities, together, constitute our ethical approach. The projects described in this book all have dedicated efforts, measurements, and assessments to make them verifiable, controllable, and accountable.

8.2 Human Versus Non-human

In harvesting data and information from the city, we come across expression by human beings and also originating from other types of entities.

This may seem clear when we deal with data coming from sensors, services generating data (e.g., a system which measures credit cards transactions in a certain area, to expose patterns of consumption), or with systems which operate on infrastructures such as highways, communication networks, and sewage.

We would expect them much less, for example, on social networks.

We would be wrong.

One of the most constant twitter users found in the Kansas City case study in Chap. 7 was Kansas City Weather (@KansasCityWX on Twitter), a bot, an automatic software which publishes on social networks depending on the data generated by a service which measures weather conditions in the area. Of course, this specific Twitter account was excluded through the filtering processes, because it is not an innovator, and it does not see Kansas City as an attractor for innovation. But that is not the point.

The point is that there are millions of expressions which are not generated by people, but by other types of entities, ranging from organizations communicating their latest news, financial services describing market condition updates, earthquake sensing and alerting services, news agencies, and many, many more types.

Even if the average human being is able to spot them in an instant, they are far from being neutral in Relational Ecosystems.

Imagine if the earthquake sensing station would publish a serious alert in your city: It would be caught in the news, spread through the digital grapevine, and propagated to emergency services, and you could find yourself moving out of your home.

Non-human entities expressing in the digital realm—whether it is through social networks, sensors, devices, services, organizations, or other means—fully participate to the Relational Ecosystem, influencing people's behaviors, imagination, actions, relations, opinions, emotions, and expectations.

We will dedicate an entire chapter to this fact (Chap. 11).

8.3 Geography of Relations

While some efforts have been dedicated to highlighting the effects of geography on the manifestations of relationships mediated by technologies (Liben-Nowell et al. 2005; Escher 2007; Wellman 1996), a smaller corpus of research describes goes in the opposite direction, analyzing how digitally established relationships affect physical locations. Even less explored are the ways in which digital and physical mutually influence each other, and the resulting potential of revealing insights into hidden patterns behind communities.

As we have described and seen in the case studies, relations in digital domains extend well beyond administrative boundaries or geographical bounding boxes, through relations which are created directly, indirectly, by sharing knowledge and information, through the actions of algorithms which establish the fact that certain users may be related to some other ones, visually highlighting their content, and in many more ways.

From our perspective, it is important to study these relationships beyond administrative boundaries, to understand whether networks moderate spatial effects or enhance them; whether they moderate or disintermediate them; whether geographies can become the condition for the formation of networks, and whether they can moderate network effects, in desired or undesired ways. And we are interested in understanding possible ways in which the insights gathered in these observations can be used practically, to use geography, networks and digital environments, and hybrid contexts to promote inclusive participatory processes, in which the digital and the physical are transmedial factors in the same equation of inclusive collaboration.

In this sense, the relations emerging from the Third Infoscape constitute a layer of the landscape which is interconnected with the physical environment, but which has completely different spatial and temporal characteristics: Its space is nonlinear; its time is nonlinear and asynchronous or synchronous, depending on the type of access and interface modality. Both can provide different perspectives of each other, without the assumption for completeness or truth, and the awareness of the possibility for coexistence of multiple interpretations of the same phenomena, according to the personal, cultural, spatial, networked, noisy, dusty, pulverous points of view, and of the possibility for constant, complex recombination of all the elements involved.

In this sense, geography also needs to benefit from a new aesthetic, to appreciate its benefits and potential beauty. We are culturally accustomed to contiguous spatial representations: maps, paths, distances, and routes. What does a nonlinear geography look like? What does a map in which someone from Rome sits next to someone from Colombia, because they influence each other in regard to what movie to see at the cinema?

And establishing once again a parallel with Clément's Third Landscape: "what does a garden without a form look like?"

We find ourselves in a similar situation, as the current intention of the concept of form is not able anymore to explain our world, to protect us from the unknown.

A new aesthetic must be invented for geography, in which "space" is defined and represented through its physical, psychological, informational, relational, temporal, and emotional character, through flows, interconnections, and remixes.

To define this new aesthetic or, better, to construct and describe how this new geography is represented and perceived, we need to describe its physics and dynamics.

8.4 Relations in the Third Infoscape

According Bateson's (1972) approach, consciousness can be observed in the ways in which it evolves and forms patterns both at social and individual levels. In the Ecology of Mind, consciousness, or mind, is modeled as an ecosystem, in which ideas are subject to evolution, extinction, or transformation. According to Bateson, by adopting this perspective we become able to describe constructively the form of the ideas produced by our society, understanding how information, ideas, and knowledge flow and are processed in the interactions with one another and across individuals and their relational networks.

In this scenario, consciousness is not a closed system.

Throughout the Steps to an Ecology of Mind, Bateson expresses this concept by highlighting the need to understand our functioning within the ecosystem (natural, psychological, social, and political), using this method to understand not only natural interactions, but also social and psychological ones, pursuing this understanding not to achieve power (and its architectures), but beauty.

In this process, he says, we can obtain forms of wisdom which can help us shape the changes in attitude that we wish to see happening in the world.

To do this, we must reconsider the boundaries between ourselves as individuals, our relations, our environment, cities, and societies.

It was since the Process Philosophy of Alfred North Whitehead, from which Bateson drawn upon, that it was noted how the fundamental unit of evolution was not the organism, but the organism together with its environment. This was also the thought expressed by cyberneticists Humberto Maturana and Francisco Varela, which observed how organisms were structurally coupled and coevolving with their environments. A notion, for example, will find deep implications in the evolution of human beings in their technological and information environment, with Castells.

Here, the role of the relational ecology is complex. Haeckel describes it as the networks of internal and external flows that operate at multiple organizational scales or orders. They are organism–environment relations which include, in the case of human beings, social, cultural, and economic agencies.

It might be observed how the basic definition of ecology expresses the desire to study the relations between organisms and the environment, arriving to its shared concerns with urbanism and architecture, which may be interpreted, in this sense, as the production of the environment by human organisms.

Among the most interesting innovations brought on by Bateson to the concept of ecology is the idea that the quantitative focus on energy and material flows was problematic. First of all because ecosystems must be considered as communicating informational systems, not just as material and energetic ones.

And, second, by making the point that we must recognize ourselves as part of the system being observed, in order to be able to think ecologically and to understand ecosystems.

As Harries-Jones (1995) has noted:

> "*Bateson realised far ahead of his contemporaries that the primary source of error in ecological science lay in false presumptions of an ability to 'control' and 'manage' ecosystems through quantitative measurement.*"

In this framework, ecosystems are ecologies of mind which are constructed relationally, in networks, through their activity and life processes, as expressed by flows of information, emotion, and opinion.

According to Bateson, to grasp this concept and its implications, it is not sufficient to just transform our knowledge, reforming it into linked, relational knowledge. A transformation of science in general is needed, creating an aesthetic dimension, a form of recognizable beauty which is stimulated by the patterns in interconnectedness within ecosystems.

In his words:

> "*So by 'aesthetics' I mean responsiveness to the pattern which connects. The pattern which connects is a meta-pattern. It is a pattern of patterns. It is that meta-pattern which defines the vast generalisation that indeed it is patterns which connect.*" (1977)

All of these considerations contribute in describing a model in which human and non-human ecosystems (e.g., communities, cities, organizations, and companies) can be observed as evolving networks and as the flows of information and knowledge which traverse them. In this context, a concept of aesthetics may be promoted and pursued, describing the possibility to reveal and perceive the emergence of patterns and meta-patterns in which networks configure themselves and information and knowledge flow through them. Acquiring a sensibility for these aesthetics grants us the opportunity to learn how to better understand these ecosystems.

This brings us back to our initial question: How can this formulation lead us to constructive action? How can this description of ecosystems become the incipit for learning, for allowing these aesthetics to become novel forms of open knowledge production and access?

In short: How can we use this view on human and non-human ecosystems to produce usable, meaningful, and beautiful knowledge, as individuals and as members of communities and societies?

There are a number of approaches which, over the years, have confronted with this type of questions.

One of them is the Actor–Network Theory (ANT), first suggested by Michael Callon, Bruno Latour, John Law, and others, within the Science and Technology Studies.

ANT is commonly described as a constructivist approach which tries to explain events and innovations in terms of the combinations and interactions which brought to them. It describes environments as social networks (both human and non-human, meaning networks of entities of different kinds which have social interactions among them) in which, for example, the success of a certain innovation is not described quantitatively, or in terms of true or false, but according to the network interactions which brought it into emergence.

This is a very interesting point of view, as the process of evaluation of a certain event or innovation (e.g., the production of a bit of knowledge) is not seen as a quantitative analysis, but, rather, as the analysis of the network patterns (configurations, relations, interactions, and flows) which brought to it. ANT does not explain how or why a network takes the form that it does. In Latour's terms, "explanation does not follow from description; it is description taken that much further" (Latour 1999).

In ANT, the term network is both fitting and problematic, at the same time.

It is fitting, because it represents systems which has "no a priori order relations," and "it is not tied to the axiological myth of a top and of a bottom of society; it makes absolutely no assumption whether a specific locus is macro- or micro-" (Latour 1996). It is similar to Deleuze's and Guattari's rhizomes.

At the same time, the term network is problematic for its unwanted connotations: first of all, because the processes observed by ANT do not necessarily happen in a network; second, because it implies the concept of transmission without deformation, which, according to Latour, cannot happen, because any actor-network implies flows of information and knowledge from one actor to the other under the form of a series of translations.

Each entity (actor, rhizome) translates information in order to understand it (or in order for the other to understand it), necessarily transforming the message.

It is a material-semiotic approach, in which material-semiotic networks are understood as a whole.

The objective is not to understand reality (or sociality) under the form of static, set, defined contexts, but under the form of liquid transformations, translations, and reassemblages of their linkages: "social, for ANT, is the name of a type of momentary association which is characterized by the way it gathers together into new shapes" (Latour 2005).

It is, thus, the science of selecting the strategies according to which these relations can be observed in their emergence, life, and extinction, as they are transient and they need to constantly be performed to keep networks from dissolving.

According to ANT, social relations are constantly in process, and they need to be performed to stay in existence.

Actors, from this perspective, are both human and non-human, and they assume different roles in networks according to the form of their relations with one another.

This may seem strange, but it constitutes our common experience of daily life. While some socially relevant transformations can happen by interacting with some other human beings, some others may easily emerge because of actions (performances) by non-human entities: A company puts out a certain job advertisement; a new vaccine is being marketed; a new law is voted; a new catastrophe occurs.

In all of these events, organizations and processes assume their place in the relational network and start living in it, from their birth—as interactions begin—to their extinction—because they are not performed anymore, because no rhizomes in the network perform their relations with them.

Actors are mediators of transformations happening in the network, and they transport (they circulate, as we have said in earlier chapters) transformations from one point of the network to the other (or others), through performance: "The word 'translation' now takes on a somewhat specialized meaning—a relation that does not transport causality but induces two mediators into coexisting."

ANT is also defined as "sociology of translation" in this sense: "There is no society, no social realm, and no social ties, but there exist translations between mediators that may generate traceable associations."

All of these items depict a vision of the network as the description of a series of actors which perform (translate) transformations to a network by making and unmaking associations (relations).

In Latour's words:

> "*a good ANT account is a narrative or a description or a proposition where all the actors do something and don't just sit there. Instead of simply transporting effects without transforming them, each of the points in the text may become a bifurcation, an event, or the origin of a new translation. As soon as actors are treated not as intermediaries but as mediators, they render the movement of the social visible to the reader. Thus, through many textual inventions, the social may become again a circulating entity that is no longer composed of the stale assemblage of what passed earlier as being part of society.*"

Action and performance come under the form of expression: through text, gestures, motions, and traversals and through the creation, modification, and destruction of relations.

Expressions do not have a definite scale or granularity: They may range from micro (one post on twitter), macro (one major catastrophe happening somewhere on the planet), and hybrid (one important law and the microinteractions through which it resonates).

Expressions may be described as statements, either in simple ways (such as texts) or as more complex ones: A company releasing a new product is making a statement not only by releasing it on the market, but also by declaring the organization's view on the world, its strategy, its forms, and modalities for communication,; the same can be said about a human gesture, or about one's dress code in relation to a certain event or network configuration, and through which we actually make readable statements about our performance, relations, and beliefs.

Latour says that "collecting statements not only traces new connections but also offers new highly elaborated theories of what it is to connect. They perform the social in all practical ways."

Each expression/statement also provides second-order descriptions about how the social worlds should be formatted. As soon as these statements can be collected and observed as polyphonic expressions "circulating along tiny metrological chains"—so that the chains and their transformations become the object of measuring—"they clearly become the source of what we mean by being in a society."

This possibility to recognize the forms of these connections, relations, and translations constitutes the opportunity for the emergence of a new aesthetics, a new perception, of the ways in which information, knowledge, innovation, and change are produced.

Talking about the second-order empiricism promoted by ANT, Latour explains how "its science, its politics, its esthetics, its morality are all different from the past. It is still real and objective, but it is livelier, more talkative, active, pluralistic, and more mediated than the other."

8.5 Roles

The idea of describing roles could be helpful in gaining insights about the form of connections within the networks in the Relational Ecosystem.

From what we have seen, roles should be described as recipes, as recurrent patterns, but with a deal of flexibility. Roles should denote recurring configuration of network flows which enable observers to obtain understandings about how they happen, and of the evolving architectures which lead to contagion, propagation, influence, and transformation.

These roles should not be exact, meaning quantified. They should allow recognizing a form, not understand whether a certain node communicated n times with a certain other one. Recognizing roles should be a similar activity to recognizing plants in a wood: You have to look at the leaves, at the stems, and at the fruit and take in consideration the habitat. Because many plants have similar leaves and fruits, some leaves could be broken, or only partially visible or accessible; they could be in different states of maturation; they could be smaller or larger, depending on how much nutriment they find in the soil and on the condition of the weather; other plants could have grown beside them, forcing their form into unusual configurations because of spatial or other contextual reasons; shadow might have grown where once there was the sun, making the plant twist and bend to arrive to the light, and so on.

Roles as we need them, should have these characteristics: They should emerge from a contextual, holistic, aesthetic observation and understanding of the context.

On top of that, roles should also be contextual in another sense: A node in the network could have different roles depending on which parts of the network you observe.

Let us make a practical example, using the diagrams shown in Fig. 8.1.

In the image, we can see a generic network, in a peculiar condition. At a certain point in time, in fact, the network is composed of four distinguishable subnetworks

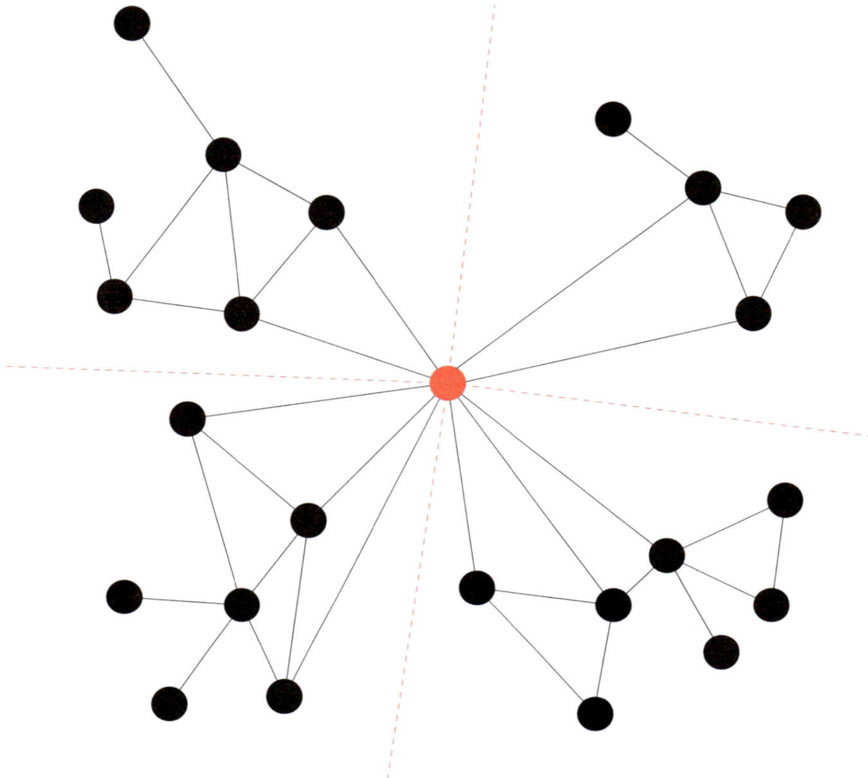

Fig. 8.1 A node can have different roles in the network

(in the image they are delimited by a red dashed line): Each one of them is well connected among its networks, but not with the other subnetworks. The only exception is the red central node, which belongs to all the networks: The red node is the only point of connection between all of the subnetworks.

This might seem as an unusual situation. But it is not. Or, better, it is unusual in this version, whose characteristics have been exaggerated to make the example clearer. Similar, although more relaxed, configurations happen all the time, from social networks to condo apartment buildings, to offices, schools, and more. People have different interests, desires, and tastes.

Imagine if the image was representing the condition of a certain user on a social network. The user could be passionate about Japanese noise music, South American cuisine, chemistry, and artificial intelligence. For this, he could join four interest groups, one for each of these topics. Now, these four topics are far enough to include the possibility that, for example, none of the users of the South American cuisine are also members of the Japanese noise music community, except for our red user. Similar situations could happen for all of the other groups. This would lead to a situation which is similar to the one in the image: The red user is the only point of connection among these recognizable subnetworks.

The red user might have different roles in each of these subnetworks. He could be a fan of a particular band in the noise music group, waiting for their updates. He could be an amateur chef, grabbing hold of all of the recipes in the South American cuisine group, try them out, and advertise on his achievements inside and outside of the group, through wonderful images of the dishes he prepared, and also by providing suggestions on how to make them come out better. He could be an expert researcher in chemistry, providing competent information to other participants in the chemistry community. He could be interested on how to merge artificial intelligence with chemistry, to discover new compounds and molecules automatically, and look in the artificial intelligence subnetwork for connections, maybe also bringing them over to the chemistry group.

All of these roles (some of which we will define in the next sections of this chapter) have no problem in coexisting in the same node of the network. They just regard different parts of the network, in a specific space/time.

The spatial component is not really evident in this example, but it may become so if we change the example to a physical one, instead of networks: Let us try with a condominium, an apartment building. The relational network could represent, in this case, the relationships within the building. The four subnetworks, for example, could represent the fact that the building has some issues to be maintained: The elevator has stopped working; there is a leak on the second floor; the roof needs to be repainted; the lights went out in the lobby of the fourth floor. The various people in the building could be focused on specific issues. The people at the last floor would be more focused on the problems with the elevator, because they have a hard time getting home with their groceries; on the other hand, they could care relatively little about the leak on the second floor, maybe even to the point that they do not discuss about it at all. Similar considerations could be done for the various other issues, giving rise to the four subnetworks: Everyone is only speaking about what they care about. And this scenario would really highlight strong spatial factors driving these relations: In an individualistic scenario—as are most condo apartment buildings—only the ones whose ceiling matches the leak would intensely discuss the water leak, the others would tendentially ignore it, or even not know about it. In this case, who could the red dot be? The building manager? Maybe. Also in this case, he would be the only node in common among all of the subnetworks, across space and time and topics (and also emotions, if you have any idea about the intense emotions which manifest themselves in regard to building maintenance, and the fact that "I have to pay for fixing the elevator even if I live on the ground floor").

The situation of our generic network could mutate dramatically with the addition of a new contextual element, as pictured in Fig. 8.2.

What would happen, for example, if the red node on social networks from the previous image discovered a new group (e.g., on Facebook) where the connections between chemistry and artificial intelligence were discussed, and joined it? Probably, he would also advertise it on the other two subnetworks (the artificial

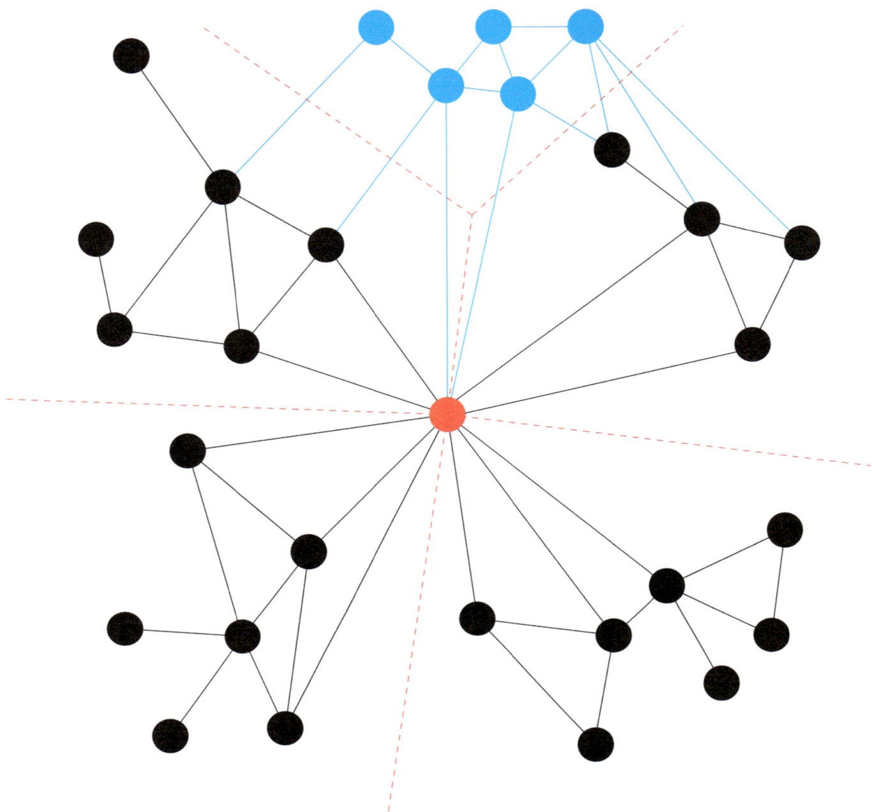

Fig. 8.2 Network scenario changes when a new element is introduced

intelligence and chemistry ones), to provide useful information, to increase his reputation capital (by doing so, in fact, he would stand out in the groups, by providing useful information), and for other reasons, mainly because he cares about this subject, because it falls within his interests and curiosities. What would happen is that some of the other nodes would join in, too (especially if the red node had some influence and reputation to begin with), and participate to the discussions in the new group, and also bring some of them in the previously existing ones.

Through a simple discovery, the scenario would immediately and completely change: The red node would not be the only point of connection between two subnetworks; discussions would expand, in unexpected directions; new ideas could be generated; the roles could change, as new subjects would enter the scene in new modalities. Within a few months from this transformation, entirely new research projects could be set up.

This, by the way, is a perfect example of Digital Urban Acupuncture: A simple, localized, transformation (discovery of a new community, which the digital acupuncturist could have performed by just introducing the red node to the new group) radically changes the scenario. We will see more about that the next chapter.

To explain how these mutations would tendentially happen, we would need a better understanding of the participants' roles within the ecosystem, also within the subnetworks which they are part of at a certain point in time. We would, again, refactor this analysis after the mutation took place, to see "who became who," and to observe their roles' evolutions.

To perform this type of task (understanding roles), we need a few assets:

1. The description of the Relational Ecosystem, in terms of a *network graph* (nodes and edges);
2. Sufficient *metadata* for each node and edge which are part of the Relational Ecosystem;
3. An *objective*, describing the ways in which we aggregate the subnetworks within the ecosystem;
4. A *scope*, including considerations in time, space, and granularity of the observations and deductions;
5. A *discovery and sensing mechanism*, to sense the transformations as they happen, in emergent ways, both within and outside of the Relational Ecosystem, and to make adjustments, including other nodes and networks in the ecosystem, excluding them (as they stop performing, as noted by Latour), or even mutating the ways in which we aggregate things together (the strategies), or our objectives and scopes (which may have become meaningless), needing for other scopes and objectives.

These requirements are shown in Fig. 8.3.

As we have seen in the previous chapter, given the proper technical, technological, and legal frameworks, and a relatively small level of initial engagement on the side of citizens and organizations, it is possible to bring up and start observing the Relational Ecosystem (may be by bringing up an instance of the Human Ecosystems and Ubiquitous Commons).

The metadata which gets captured and generated in this process should be relevant and meaningful in relation to the scopes and objectives which we set forth. How do we know that? How can we make sure that it happens? Obviously, there are no sure ways. But there are good practices and approaches which can be adopted. On the one hand is strategy, and on the other hand is the sensibility to emergent tactics.

In regard to strategy, it must be said that these kinds of processes (e.g., bringing up an Human Ecosystems instance) do not happen in a vacuum: They cost money, resources, maintenance, and management. For this reason, they usually find their origin in something which is already happening in the city: an event; an initiative by the public administration; and a call for proposals; all regarding issues which can vary considerably, such as energy, public safety, citizen engagement, traffic and

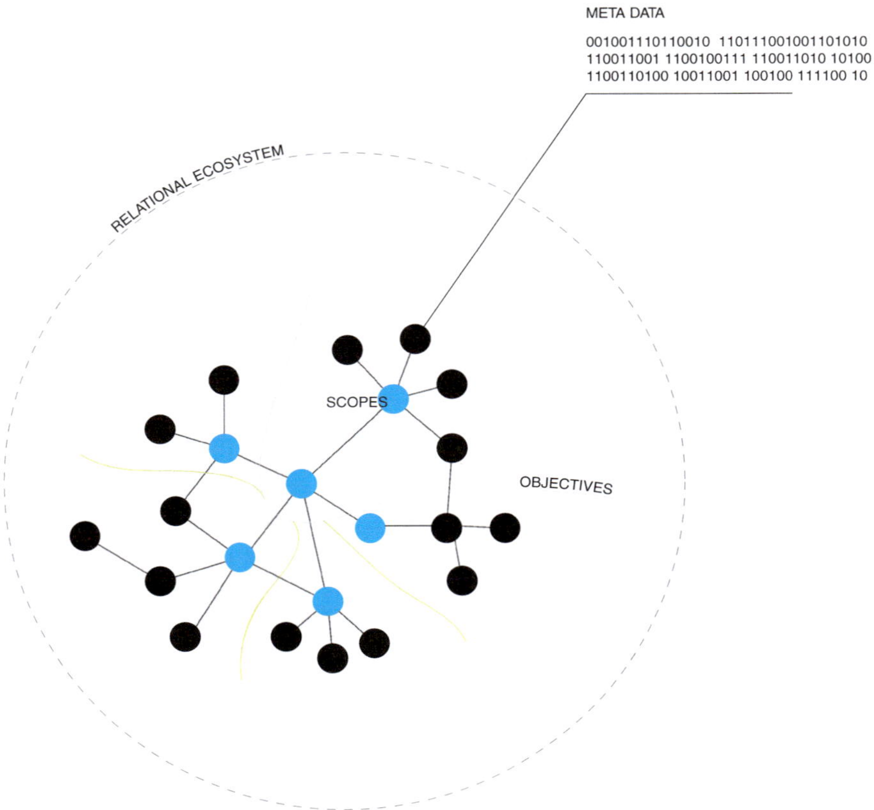

Fig. 8.3 Infrastructure and requirements for discovering roles in the Relational Ecosystem

mobility, education, innovation, and more. Strategies are formed in a top-down fashion, and thus, their initiators are fairly easy to recognize and contact, to propose the application of this methodology: Digital Urban Acupuncture, or also starting from more limited, scoped, processes, such as simply bringing up a Human Ecosystems—or other similar—instance, and including in the offering enhancements directed to the creation of cultural instances in the city, to create engagement and innovation (in our approach, these may be driven by the Real-Time Museum of the City, and through the Laboratory which enacts the inclusive education programs).

In regard to tactics, it involves getting in touch with the bottom-up scenario of the city, contacting grassroots organizations, citizen groups, cultural initiatives, single events, and more, to bring the approach into the city and causing contagion in the city about the effectiveness of the approach and about the benefits which can come by its adoption. In this case, action needs to happen at the level of the

Relational Ecosystem of the city, since its beginning, as part of a participatory performance in which the digital urban acupuncturist and the citizen collaborate on addressing their needs, goals, desires, and visions.

In both cases, the initial set of metadata which will characterize the harvested information will be limited: Both types of approaches, in fact, imply that the observation of the city will begin by focusing on a set of predefined topics, be them mobility, energy, water, safety, policy making, or else. In fact, we have seen this in practice through the previous case studies.

In Chaps. 10 and 12, we will see the cases of Rome and Sao Paulo, which show how, starting from a definite set of topics to be observed in the city, spontaneous and tactical usages claimed their space in the initiatives and became actual processes in their own dignity.

In general, the metadata that we collect or generate from the harvested data will need to be sufficiently expressive to support the needs of the established scopes and strategies, and expansible enough, so that additional scopes and strategies can be defined in progress, as time goes by and situations evolve.

For example, in the Human Ecosystems, each node and relation is characterized by a unique identificator which can be associated with multiple types of data (its metadata).

For example: if, in the beginning, we could start by capturing citizens' expressions on traffic (as we could want to use this approach to enhance peer-to-peer mobility practices in the city), a certain event or scenario could suggest that we could use the same technique to also observe energy-related issues. We could, then, start capturing those expressions, too, by simply turning on additional harvesting processes and by associating the additional type of data to the identifiers of the nodes and edges of the Relational Ecosystem. It would be, to all effects, just as in the earlier example, when the red node discovered the new subnetwork.

This has two advantages:

1. The possibility for progressive enhancement and enrichment of the network and of the characterization of its nodes and edges;
2. The possibility, as we add more modalities to our observation, to discover new types of relations and flows.

In regard to item 2, for example, when we "switch on" the observation themed around energy, we might discover that it is not separate from the one of mobility, just as in the earlier example we discovered that the "artificial intelligence" and "chemistry" topics were not disjoint.

From the things which we have described here, it is becoming clearer how the availability of metadata supports the scopes and objectives we set forth in our observation of the city, and how they reflect in the Relational Ecosystem.

The scope is the extent of the area or subject matter we want to deal with. The objective is the goal we set forth. The scope describes "how deep" we want to dig in our investigation. The objective describes its "reason," what we want to achieve.

The scope will determine the granularity with which we want to confront with the Relational Ecosystem. Are we interested in general behaviors? In single communities? In orientations and factions within these communities? In the micropolitics which determine these orientations? In single individuals?

As a rule of thumb, in the Digital Urban Acupuncture approach, we aim at these last two levels of granularity: the micropolitics and the single subjects, which are the drivers of the flows. But other forms of investigation are possible: For example, we might wish to investigate how different communities influence each other or how certain organizations are able to confront with specific groups. Or the other way around, describing how certain communities are able to influence certain organizations or groups. The constant pattern is that we focus on the network flows, on how information, opinion, and knowledge are mobilized across the networks, in relation to the things that happen, with the consideration that we can do that at different granularities.

The objectives will determine what metadata we capture and how we aggregate nodes and relations depending on it, by creating clusters. Given a certain objective (let us say "mobility"), we will tend to collect data and information which is relevant to it (e.g., people's expression on social networks about mobility; their reactions to certain news items; the data from cars and traffic; people's movements in the city to go to school, work, shopping, entertainment, free time). We will organize this information in variables (e.g., if we record people's emotional expressions about public transportation in 5 different ways, these would be at least 5 different variables), for example describing the fact that they have joined a certain group on Facebook (a boolean value, expressed in terms of true/false), that they share the urban paths they do with their bicycle (number of kilometers, geographical indications of the paths, and their distribution over time), and that they express on certain public transportation relevant topics (the list of topics, the time sequence for their emotional expressions on these topics).

This together with the explicit or implicit relations which they express (explicitly, as they comment on something, retweet it, share it, provide information and answers, appreciate it; in general, from all of the things, we can learn through their public data activity on social networks, devices, sensors, domotics, Internet of Things devices, etc., implicitly, as they assume behaviors which are shared in the dyadic relation, even if they do it independently from one another, like sharing the same link, expressing the same opinion, or other modalities) allows us to construct the network architecture (and to evolve it as time goes by, to reflect the mutating relational environment) and to characterize each node and link in the network through their relevant metadata.

It becomes possible, at this point, to use standard network clustering algorithms to classify and represent the network and its elements in ways which are meaningful for our objectives. Obviously, as the situation in the network evolves (because people join new communities, start new discussions, express new behaviors and more), these algorithms should be applied again, obtaining a new description of the network and constituting a time series of the evolution of the network over time.

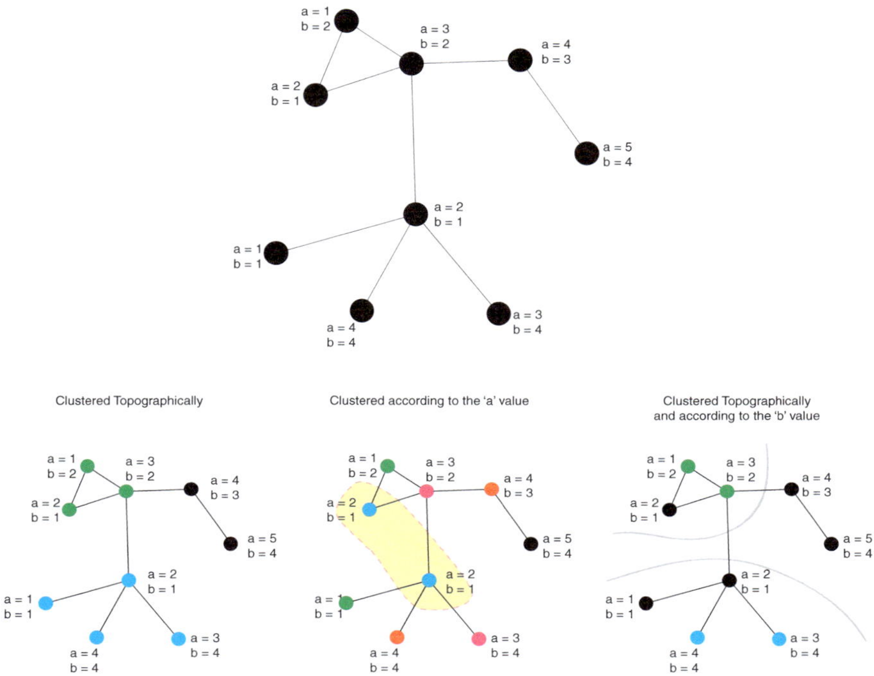

Fig. 8.4 A simple network, clustered in different ways

Clustering can be performed in many ways, mainly two:

1. By considering the topology of the network;
2. By considering the metadata of the network's elements.

Each method has its uses.[2]

Both methods can be combined with one another (e.g., using a topography-based mechanism to identify communities, and within them, a metadata-based algorithm could be used to discover orientations, factions, similarities, differences, and more).

Figure 8.4, for example, shows a simple network and how it would be interpreted through different clustering strategies. A topographical clustering would determine those parts of the network which are tightly connected. The green network in this first diagram on the left has, in fact, all the nodes connected to each other: It is a tight community in this topographical determination. In the second

[2]The Human Ecosystems software platform uses and combines both approaches, for different purposes. Readers are encouraged to download the software, install it and use it, to practically learn the usages of these techniques, and the effectiveness of their results. To download Human Ecosystems refer to the notes in the previous chapters.

diagram, at the center, clustering is done through one of the variables which are associated with each node: variable "a." The result is that clusters are formed wherever "a" has the same value. Note, for example, the cluster highlighted in yellow and surrounded by the red dashed line: "a" has value 2, and it becomes really easy to understand which communities this cluster is a part of, and to what is related (just follow the links which cross the red border). The third clustering is a local one, meaning that a topographic clustering is performed, and then nodes with a certain value (for "*b*" in the example) are clustered together. It means that same values for the variable are searched for only within the cluster, not beyond it. It is local. Note, for example, how the "$b = 4$" nodes in the lower topographical cluster are blue while, instead, the one on the right is still black. This is useful to find modalities, factions, and trends within each cluster independently.

Clustering is the first step in understanding nodes' roles in the networks. It allows us to understand:

- to which cluster (topographical or value-based) the node participates to;
- what are its behaviors in relation to the interior and to the exterior of the clusters;
- the variations in both points over time, space, and context.

This knowledge allows us to look for patterns, recurring forms, to be intended as a mathematical attractor, a tendency in which the node gravitates, which it seems to respect in coherent ways: a form which tends to be maintained. It is the form which we were looking for.

8.6 A Selection of Meaningful Roles

In this section, we will present a selection of meaningful roles which may become manifest within networks.

As said, their definitions should be taken with extreme care, as they are the *sui generis* description of forms: They are tendencies, behaviors which tend to be performed in consistent ways, give or take some differences.

Not every node in a network has a role of these kinds. Nevertheless, it is possible to discover new recurring patterns which might be of interest for our research and, thus, define new types of roles, as needed.

These roles listed here come from our own experience and describe meaningful patterns which have helped us explain how networks behaved.

In the discussion for each role, parameters are mentioned such as horizontality, verticality, weight, persistence, and others: They are the ones we introduced in the previous chapters to characterize relations.

(a) The expert

Figure 8.5 shows a diagrammatic representation of an user behaving as an expert.

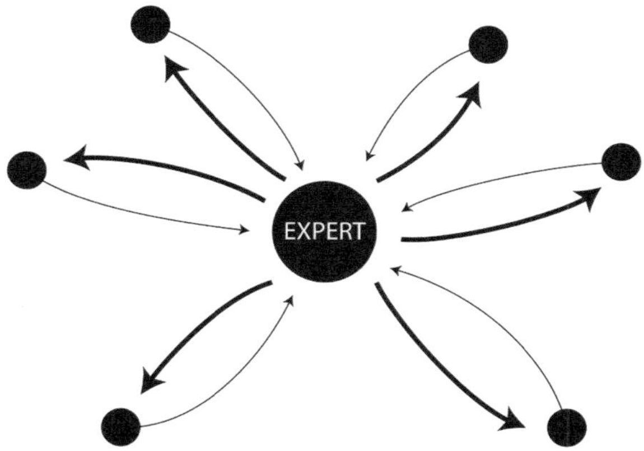

Fig. 8.5 The expert

The expert has many relations, often of the vertical type (the expert's field of expertise). Multiple nodes, often across clusters (both topographical and value-based), refer to the expert for information. This fact can be described by using, for example, the natural language analysis features of the systems we use to capture the data (e.g., by using Human Ecosystems). For the expert, many of these requests and referrals will come under the form "do you know X?", "what do you think about Y?", "is Z true?", and similar ones.[3]

The expert's outgoing communication and information (the arrows going out from the center) have a large weight: They often stimulate further discussions, generate comments, and polarize among different factions. They quickly establish influence and relevance for the people who receive them (meaning that they are often quickly satisfied with the answers, so much that they do not pose further questions of the same kind, or they even share the answers, mentioning the subject who provided them, the expert).

In synthesis, the expert does what experts are commonly expected to do: If asked, they provide highly relevant answers, which are used in further interactions and which often generate polarized responses, divided among who agrees and who does not agree with the expert's opinion.

An observation of this kind could be performed by defining a few variables associated with nodes:

- is asked/gives information on topic X (a counter, increasing each time this happens);

[3]Natural language analysis can be proficiently used for this purpose even in its simplest forms. Discourse analysis, for example, can be used in many cases by understanding the form and structure of textual exchanges, to understand whether they are questions, to which an answer is given: These tasks are performable automatically, by inspecting the grammatical structure of sentences, through their subjects, verbs, complements, punctuation, and other features.

- weight of answers on topic X (time series, with an additional value corresponding to the weight of the answer, each time the expert is asked for information on topic X);
- verticality index (an index which measures the verticality on certain topics of the relations established by the subject);
- dispersion, numerosity, and persistence of relations (if someone is recognized as an expert, he may receive requests from numerous different subjects, even from contexts which are far from his own, and the possibility for a low persistence of relations becomes higher, as people ask, and leave).

These are just a few of the variables which, explored in time and across clusters, may reveal the recurring patterns which would lead us to define the subject as an expert.

There are many non-human experts, in this sense. They often do not wait for someone to present them with a question and tend to express automatically, on their own, regardless of the fact that someone is actually listening. For example, the earthquake sensing and alerting bots are a perfect example for this: Whenever they express, a high number and variety of people take their opinions in serious, influent consideration.

(b) The hub

The diagrams shown in Fig. 8.6 show the behaviors over time of a hub.

To understand the behavior of hubs, we have to observe them over time.

The hub establishes many relations, interacting with a variety of subjects. Some of these connections have more weight than others, as the hub takes them in more serious consideration (as can be seen, e.g., by the fact that the hub chooses to highlight some of them by sharing some of their information, by retweeting, commenting, reposting, or other activities and processes). In this stage, the hubs' two types of coexisting networks are as follows: One has stronger ties and another

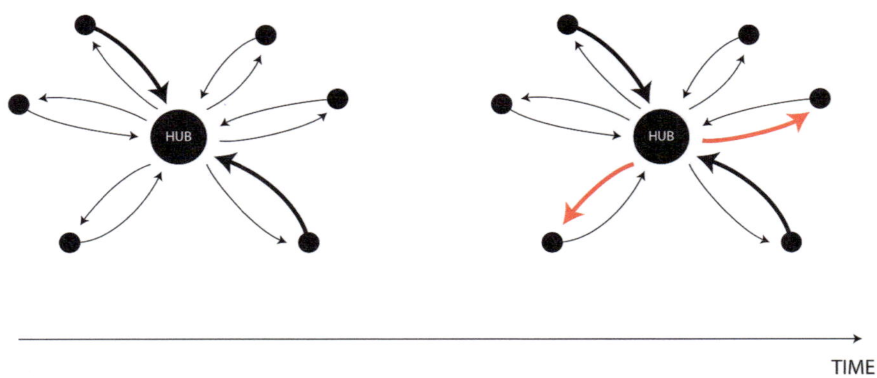

Fig. 8.6 The hub

one has weaker ties. This is the process which we have identified as the hub forming its opinion.

In following stages, the hub chooses to facilitate some connections (red arrows on the diagram on the right), putting subjects in contact. This, for example, can be performed by mentioning different subjects in the same message or by proposing themes for common reflection to multiple people.

This is a cyclic behavior and can repeat over time. When observed along the time axis, several—eventually overlapping—cycles of this kind can be recognized.

This two-phase behavior is the main characteristic of hubs: They clearly distinguish a pull behavior from a push behavior: In the former one, they attract, creating a capital of relationships; in the latter one, they disseminate, augmenting its value (by creation of reputational capital, by means of making introductions, as the simple sentence "Hey X, you should meet Y to talk about Z" may place the Hub in a strategic position.

To measure this type of behavior, a series of variables could be defined:

- the number of relations as it evolves in time;
- the correspondence between relations and their weights (which can assume a form of a matrix, whose transformation over time we could visualize as an image, with the pixels corresponding to the cells in the matrix, and in which we could recognize forms and patterns by calculating simple differences);
- the connections promoted by the subject (which could also come under the form of a matrix with the subject's relations on its columns and rows, and the resulting cells highlighted when the subject introduces the two other subjects, becoming the mediator of a triadic relation built through two dyadic ones).

(c) The influencer

Figure 8.7 shows the behavioral diagrams of an influencer.

This role, as well, needs to be observed over time.

The influencer has numerous, asymmetric, connections.

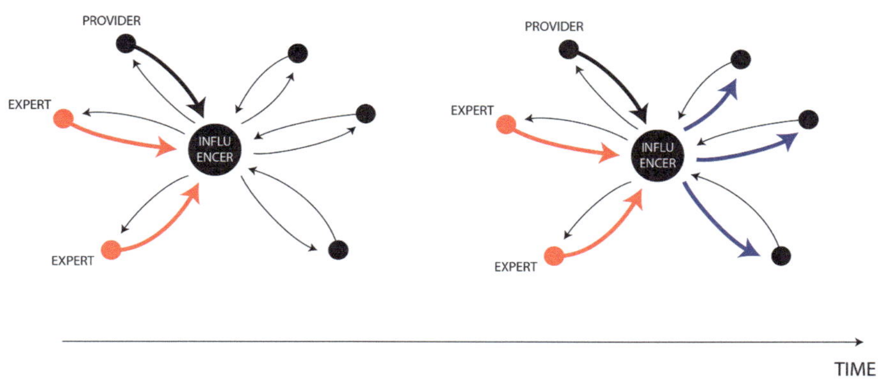

Fig. 8.7 The influencer

A limited number of these connections act as information providers: They are sources of news, information, knowledge, curiosities, and more. The influencer takes this information in serious consideration (thus their high weight). Influencers rapidly evaluate these types of connections, constantly adopting new ones, and either quickly persistently connecting to them (transforming its relationship with them into a high-weight relationship), or discarding them and passing to new ones. The influencer accurately chooses its sources (e.g., influencers often follow only a small amount of carefully selected accounts on Twitter).

Occasionally, the influencer combines the information coming from providers with the expertise provided by experts (as described in the previous sections). This can be done, for example, by mentioning experts in tweets and Facebook posts.

Whenever experts provide feedback on these forms of engagement (the red arrows on the image on the left), the influencers take these expressions into high consideration (high weight), and often forward them to a wider audience (the blue arrows on the image on the right), provoking a wide discussion and, in general, influence (affect discussions, provoke transformation in opinions and emotions, etc.).

It is possible to see influencers as specialized information and knowledge brokers: actively curating sources, selectively asking for expertise from experts, and disseminating the results. They often resemble a one-man-publishing house.

Influencers can be detected using methods which are similar to the ones we have used for experts and hubs. In fact, they are "expert hubs," with the additional character of establishing conversations with wide audiences: They are deemed relevant by many subjects. .These conversations, sometimes, may be of the broadcast type: Many influencers substantially emit information and establish few real interactions.

The main characteristic of influencers is these three phase behaviors:

- sources curation;
- experts curation;
- forwarding.

As usual, the three stages can be overlapping in time.

The list of variables which could be set up for observation to identify influencers includes all of the ones for experts and hubs with the addition of:

- numerosity of followers and reactions (retweets, shares, comments);
- The measure of the three steps of behavior is as follows:
    - sources curation, captured by the ways (numerosities, relations) in which the subject follows/unfollows sources of information, either explicitly (e.g., they follow/unfollow them on Twitter), or implicitly (they stop performing the relation);
    - experts curation, captured by how they ask for information to a changing or stable set of subjects, and what they do with the answers which result (this is also a matrix);

- forwarding, captured by the way in which they receive or capture information from sources, forward it to experts, then forward a synthesis to the "audience" (in quotes, since audiences of this type may be so active and participant that the use of this term is rather problematic).

There are many non-human influencers around. Excluding the easy category of organizations, there are multiple forms of automatic influencers. On social networks, they are often the results of news aggregators and artificial intelligences, or sometimes, they are as peculiar as "Two Headlines"[4]. Influencers are widely diffused also in other contexts, for example, from e-medicine, in which multiple services are engaging in this pattern. Among the best known are the "flu near you" types of services, which combine selected information sources with experts evaluations to create alerts about flu epidemics which are spread to vast audiences through apps, wearables, biomedical devices, and more, creating influence and reactions.

(d) The amplifier

Figure 8.8 shows how an amplifier behaves.

The amplifier is very similar to the influencer, but with the influencers replacing the experts, as shown in the image. This provokes a radical transformation, in the fact that amplifiers provide a level of curation which is much lower (it is provided, after all, by the Influencers) and they "adopt" the opinions of selected influencers and their providers of information and expertise, and make it their mission to amplify their network reach. Successful amplifiers are very useful for "getting your message out there," given you are an influencer.

(e) The bridge

Figure 8.9 shows the network configuration of a bridge.

As the name implies, the bridge is found in between two or more contexts, which it interconnects. Formally, the bridge sits in between two networks (two communities, clusters, or other relational networks of the various possible types, be them vertical, horizontal, transversal, or else).

Bridges can be mono-directional (they have influence in one direction, but not in the other), bidirectional (they have influence in both directions), or a-directional (they effectively sit in the middle of two networks, but they do not have influence on any of the two sides).

---

[4]Created by bot auteur Darius Kazemi, Two Headlines works by taking headlines from Google News, and swapping out the key noun with a different trending topic. The best tweets read like reports from a world "where there is no discernible difference between corporations, nations, sports teams, brands, and celebrities", Kazemi says. "It is generating jokes about the future: a very specific future dictated by what a Google algorithm believes is important about humans and our affairs."

While being a peculiar example, it perfectly respects the influencer pattern we have just described, including the constant curation of sources and experts (although the "experts" in this case, are other non-human experts, like Google's trending topics.

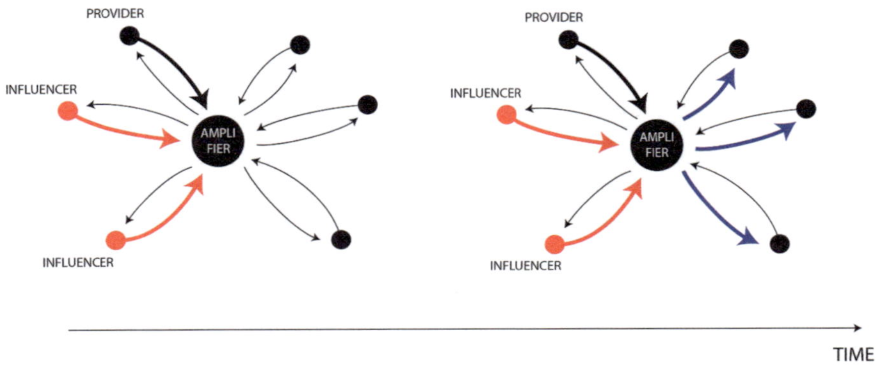

Fig. 8.8 The amplifier

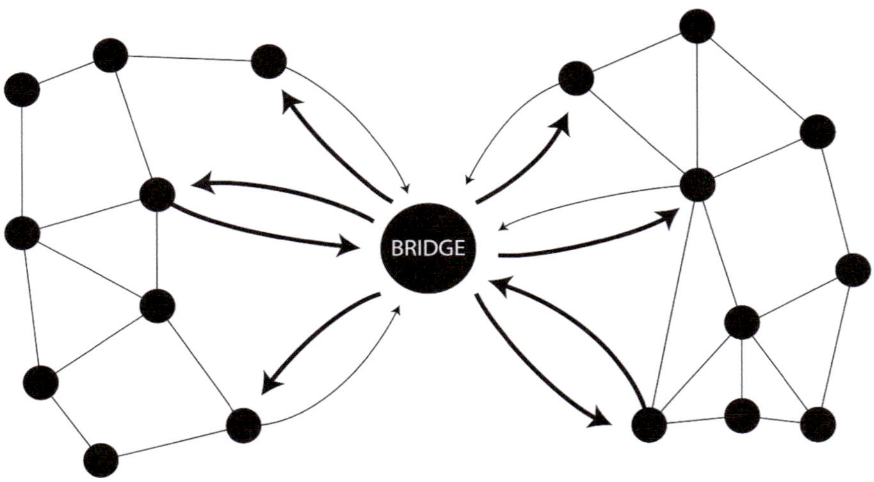

Fig. 8.9 The bridge

Bridges are among the most important tools to create engagement and activation in wide communities: They represent (in practice, or in potential, as in the case of the a-directional bridges) the opportunity to interconnect communities and cultures.

They are easily identified by exploring their positioning within the topography of networks, in their centrality, their capacity in providing connectivity to remote parts of the network, their uniqueness in this, and other topological considerations.

Through these (and others; and the others we could define when we encounter them, using the same method) roles, we are able to describe a wide variety of flows of information, knowledge, communication, opinion, and emotion across networks, especially relational networks like the ones which we have discussed so far.

We can imagine the nodes of these networks assuming different roles in different contexts: Maybe they are influencers in a certain network; experts in others; simple nodes in some, amplifiers in a few; and bridges on yet other ones. Which dynamics arise from these situations? How can we model them? How can we use them to suggest and facilitate the emergence of inclusive, participatory practices in our cities?

We will discuss just this in the next chapter, in which we will define Digital Urban Acupuncture.

References

Bateson G (1972) Steps to an ecology of mind: collected essays in anthropology, psychiatry, evolution, and epistemology. University of Chicago Press, Chicago

Bateson G (1977) Box 6 manuscripts 'Mind in Nature', Nov. 17th 1977 (unpublished), quoted in Jones PH, Hoffmeyer J (ed) (2008) Gregory Bateson's 'Uncovery' of ecological aesthetics. A legacy for living systems—Gregory Bateson as a precursor to biosemiotics, vol 158. Springer, Copenhagen

Escher T (2007) Researching the geography of social relations: analysis of the spatial distribution of friendship networks on social network sites. In: Third international conference in e-social science

Harries-Jones P (1995) A recursive vision: ecological understanding and Gregory Bateson. University of Toronto Press, Toronto

Latour B (1996) On actor-network theory. a few clarifications plus more than a few complications. Soziale welt 47(4):369–381

Latour B (2005) Reassembling the social. An introduction to actor-network-theory. Oxford University Press, New York

Latour B, Law J (eds) (1999) Technology is society made durable. Sociology of monsters. Routledge, New York

Liben-Nowell D, Novak J, Kumar R, Raghavan P, Tomkins A (2005) Geographic routing in social networks. Natl Acad Sci 102(33):11623–11628

Wellman B (1996) Are personal communities local? A dumptarian reconsideration. Soc Netw 18 (4):347–354

Chapter 9
Digital Urban Acupuncture

Abstract In this chapter we will understand how to use the Relational Ecosystem to perform Digital Urban Acupuncture. The chapter starts off from the definition of Urban Acupuncture, and from all its supporting theories and methodologies, to then define Digital Urban Acupuncture: its goals, methodological approach, assumptions, and operative strategies. By considering the city as an ecosystem, it is possible to take into consideration the myriads of microhistories of all of its members to identify flows of information, communication, knowledge, emotion, opinion, activation, and their patterns over geography and time. These can be used to highlight the pressure points, just as in acupuncture, which can be activated to creatively provoke engagement, interconnection, relation, and the breaking of undesired loops and patterns. This allows for both a dynamic definition—and polyphonic—of the well-being of the city, and to activate the identified pressure points in creative ways, to provoke wide results, affecting wide areas of the population. A final section of this chapter explores the issues connected to the multiple forms of divide which can be present in this scenario, and possible ways in which it is possible to confront with them: digital divides, cultural divides, age divides, and more. The chapter ends with an interview with Marco Casagrande.

The introduction of Jane Jacob's "Death and Life of Great American Cities" (1961) begins with: "This book is an attack on current city planning and rebuilding."

Why is such an attack necessary?

For Jacobs, it is necessary to go beyond the notions and methods driving traditional planning, and to achieve techniques which are closer to "real life" and participatory in nature. "Streets and their sidewalks," she says, "the main public places of the city are its most vital organs." When you try to imagine a city, you tendentially imagine its streets, their safety/unsafety, their characteristics, densities, colors, sounds, and smells.

"*The first thing to understand is that the public peace–the sidewalk and street peace—of cities is not kept primarily by the police,*" but by "*an intricate, almost unconscious, network of voluntary controls and standards among the people themselves, and enforced by the people themselves.*" (Jacobs 1961)

S. Iaconesi and O. Persico, *Digital Urban Acupuncture*
DOI 10.1007/978-3-319-43403-2_9

This consideration is of particular importance in our research, as it highlights the need to better understand emergent behaviors—to foster the well-being of the city—and the ways in which relational networks form and live in urban areas, as the driver for positive transformation.

It is of even more importance when we think about the transformations of the city in the digital era and in the age of ubiquitous technologies. While, traditionally, relations in the city, among its dwellers, happened in streets and sidewalks, or focusing on them—Jacobs, for example, mentions the fundamental role played by sidewalks for people's gaze, from the windows above, establishing contact, relation and building social capital and diffused social control—now the Relational Ecosystem of cities is progressively shifting toward digital dimensions, interweaving them with the physical reality.

The cities of the future will not physically differ in significant ways from the ones we know now, and their differences laying mostly in the ways in which we will relate among one another, with nature and with objects. Ubiquitous technologies, novel forms of urban interactions, and a mutated sensibility toward the environment (whether constructed or natural) achieved through culture as well as sensors, diffused electronics and relational technologies will interconnect human beings and the places and "things" which surround them, creating novel dynamics.

Some of these interactions will be mediated—for example, by service providers—some of them will happen according to peer-to-peer modalities. Even in the case of mediated relations, tendencies show how they will mostly be perceived as environments—for example, what we can understand from current social networks—and, within these environments, will appear/be perceived as mostly peer-to-peer interactions. This type of reflection seems valid for things like social networking ("I am sending my message to person X, not to Facebook, who then sends it to person X"), for domotics ("I am sensing my smart refrigerator's status, not accessing Nest"), and to most types of those network-connected devices which are shaping and will be shaping our experience of homes, offices, schools, and cities in general.

This fact calls for further reflection on the ways in which transformation in the city can take place. Adding up the factors exposed in the previous paragraphs and chapters, it seems that peer-to-peer, distributed, diffused dynamics will assume roles which are progressively more important—if not fundamental—in bringing change to urban environments.

According to our research (Iaconesi and Persico 2014), this implies the need to adopt tactical approaches which are able to establish wide, diffused, participated practices in which citizens become able to observe the Relational Ecosystem of their city, understanding the flows of information and knowledge across cultures, contexts, social groups, and topics, in order to create shared decision-making and operative processes.

Some of these approaches can be directly derived from various forms of Tactical Urbanism.

Tactical urbanism can be defined as an approach which focuses on creating small-scale interventions which are able to inspire long-term change. In "Tactical

Urbanism: Short-term Action, Long-term Change" (Lydon et al. 2011) describe it through 5 characteristics, which are as follows:

- A deliberate, phased approach to instigating change;
- An offering of local ideas for local planning challenges;
- Short-term commitment and realistic expectations;
- Low risks, with possibly a high reward; and
- The development of social capital between citizens, and the building of organizational capacity between public/private institutions, nonprofit/NGOs, and their constituents.

This approach has multiple types of actual manifestations, ranging from the ones dedicated to boosting economic revitalization of neighborhoods, to ones focused on citizens' safety and well-being, to ones providing ways for people to connect with one another and establish relationships, all at varied spatial and temporal scales, with preference to localized ones.

All applications share a performative and participatory approach: Not only citizens are welcome to the initiatives, but they are also invited to reflect on the relational and symbolic systems which cause the need for the intervention to arise and suggest possible solutions. This modality has influenced the general debate about the involvement of informal actors in urban planning processes, also in consideration of the particular cost-effectiveness of such practices: If a low-cost pilot started with the aid and collaboration of informal partners shows a sufficient degree of potential or success, it may be scaled to permanent, full-fledged initiative, allowing for continuous experimentation, high effectiveness through iterative design, and the construction of social capital through collaboration and participation.

Performative practices bear an important role in the process. By creating unexpected, unanticipated, unforeseen interventions in urban space, a state of suspended reality is created, enabling citizens to reconsider and reframe the ways in which spaces are used and the relations which can form among people, objects, urban furniture, architectures, contexts, and situations. The performance creates a mental space which is open to possibility and which allows for temporarily breaking the status quo, the common consensual reality, and reconstructing it in novel ways. This is useful, in particular, to reframe relationships. Whether it is among citizens, objects, or spaces themselves, the creation of these temporary or spatially limited interruptions of common usage patterns—the ones to which we are used to in our daily lives and which we give for granted—allows for reconsideration of the ways in which information, knowledge, emotion, communication, and experience flow among participants, in space and time.

One particular form of Tactical Urbanism which deals in interesting ways with these issues is Urban Acupuncture.

There are multiple definitions of Urban Acupuncture.

Jaime Lerner, in the introduction to his book "Urban Acupuncture" (2014), writes: "Just as good medicine depends on the interaction between doctor and patient, successful urban planning involves triggering healthy responses within the city, probing here and there to stimulate improvements and positive reactions. Intervention is all about revitalization, an indispensable way of making an organism function and change."

This is a first possible definition for Urban Acupuncture, in which the city is seen as a living organism in which transformation is stimulated with precise, quick interventions which create new energy and demonstrate the possibilities of a space in a way that motivates others to engage with their community. Lerner does not indicate clear methodologies to perform Urban Acupuncture and in all of his writings he describes as Urban Acupuncture interventions made through arts, creativity, architecture, events, street performances, recontextualizations of spaces and objects.

Another intendment of Urban Acupuncture differs from Lerner's one because instead of focusing on sustainability, mobility, and tolerance, it explores cities' growth and how citizens can take active part in it. This is, for example, Manuel de Solà-Morales i Rubi's take on Urban Acupuncture.[1] He describes cities as possessing skins made of constructions, textures, and contrasts, and defines urban experience as the contact between bodies and this skin. In this friction, certain spaces become charged with collective character, attributing energy, meaning, transforming them into semantic material. The possibility to localize sensitive intervention spots in the urban skin is what he defines as Urban Acupuncture.

Artist Gordon Matta-Clark used art and urban performance to transform abandoned building and spaces into site-specific art installations which would bring on perceptive and operational transformation in the city's fabric. In his method, he developed a system for identifying small areas of urban decay to make interventions on the first step of Urban Acupuncture.

Architecture and Urban Design professor Nicholas de Monchaux, with his "Local Code" project,[2] has expressed an interesting take on Matta-Clark's approach, by systematizing it using cutting edge digital tools and methods, including the combination of multiple layers of digital cartography to highlight, compare, and evaluate the networks of spaces in cities such as San Francisco and New York, showing how they related and interacted, and arriving at the possibility to describe how complex systems of small-scale interventions could achieve large, citywide impacts.

Casagrande's definition is the one which matches more closely the vision which is at the center of this book, focusing on an evolved, shared, social understanding of the flows of information, knowledge, emotion, opinion, expression, and communication among citizens and the environment (whether natural, built, hybrid or in-between) to be able to identify the nodes and directions in which these flows can be operated to establish new system dynamics.

---

[1]The vision of Manuel de Solà-Morales i Rubi https://www.youtube.com/watch?v=aZ8cocbuj8E.

[2]An insightful video about the project can be found here: https://vimeo.com/8080630.

A prominent Finnish architect, environmental artist, and social theorist, Casagrande[3] is perhaps the first to systematize the concepts of Urban Acupuncture. Indeed, the core principles of Urban Acupuncture are integral to his theory and practice, and his resultant work can be seen in the built environments of Taipei and Montreal. In an interview with Laurits Elkjær in I(2010),[4] Casagrande described Urban Acupuncture as:

"crossover architectural manipulation of the collective sensuous intellect of a city. City is viewed as multi-dimensional sensitive energy organism, a living environment. Urban Acupuncture aims into a touch with this nature."

In the same interview Casagrande states:

"First you have to determine the sensitive flows of the built human environment. Based on those you have to determine the acupuncture points. The last is to determine what is the needle: architectural reaction. The key to understand this is to be present. To be truly present one has to give up, one has to be weak. To be present is the key of all art."

This kind of statement is interesting for a series of reasons. First of all, for its definition of "weakness" and, thus, the necessity to create interventions which are not "strong," but able to adapt, insinuate, and shape themselves according to the flows of energy, communication, emotion, information, and knowledge of the city, its inhabitants, nature, and objects. Second, for the idea of presence, which describes both the idea of the need for participation and of performance: to perform you have to be present; the idea of the urbanist as a scientist, able to observe the "system" from the outside disappears, replaced by the one of the performer, present in the space, the ethnographer or shaman, performing his observation/research/ritual in participatory ways. Third, the idea of the need to understand these flows to enable both of the previous concepts.

This notion of Urban Acupuncture is about sensibility. In Casagrande's Third Generation City, citizens are called to be(come) sensible to the relational networks and flows which happen in the city, to cooperate with others and with nature: it's a participatory performative state.

What happens when ubiquitous technologies enter the picture in this scenario?

9.1 Digital Urban Acupuncture: A Definition

Digital Urban Acupuncture (DUA) is a participatory urban intervention methodology and practice in which knowledge of the Relational Ecosystem of cities—including the flows of communication, information, knowledge, emotion, expression, the communities in which people gather, and the roles which people play in

---

[3]Urban Acupuncture by Marco Casagrande https://www.youtube.com/watch?v=hrmQjqgSB0s.

[4]The interview is accessible at: http://casagrandetext.blogspot.it/2010/04/laurits-elkjr-marco-casagrande-urban.html.

them—is diffused and made available to all stakeholders using artistic, creative ways—such as exhibits, museums, visualizations, performances, workshops—so that they can learn how to understand it and its implications, and how to use it to achieve their goals by using the relational networks to create operative, performative connections and, thus, (re-)establish flows to "make things happen" in the city.

Given a certain topic and set a scope (in spatial, temporal, and social terms), DUA is about creating a sensing environment for the flows of data, information, knowledge, emotion, communication among human beings, the environment, spaces, buildings, objects, and thus, of the Relational Ecosystem which is defined by the networks described by these flows; about making it available, accessible, usable, and manageable by all of the stakeholders; about creating iconic, suggestive, immersive experiences, in order for the sensing environment and its use to become desirable and performable; about letting this desire and opportunities for performance enter cultures and imaginations through an inclusive education process; and about enabling all stakeholders to use all the knowledge and imaginaries they gained in the process to generate spontaneous, tactical, punctual actions and interventions, by creating, remixing, recombining flows—connecting the dots in the Relational Ecosystem— to enact wide, inclusive outcomes.

The methodology is composed of different phases, shown in Fig. 9.1:

- sensing;
- commoning;
- public experience;
- education; and
- participatory performance.

The first phase, *sensing*, requires to design, implement, and deploy a strategy for collecting information about the Relational Ecosystem, its evolving networks, and the flows—of data, information, knowledge, emotion, opinion, expression and, in general, communication—which it hosts.

As we have seen in previous chapters, frameworks like the Human Ecosystems can be used for this purpose.

In general, the sensing system must be designed in order to be able to be as inclusive as possible, capturing (tendentially) all the manifestations of the interactions relevant to topic as it appears in the city in expected, unexpected, emergent, unforeseen, and serendipitous ways. Relation is generated in multiple ways which can be implicit or explicit, formal and informal, encoded within some reference system or not, and the objective is to be as inclusive as possible in regard to all of these modalities.

Relevant data and information in cities may manifest itself in a plurality of ways, and distributed according to multiple layers, the principal of which are shown in Fig. 9.2.

There is bureaucratic and administrative data and information, the one which is generated and managed through governments and administrations, and which

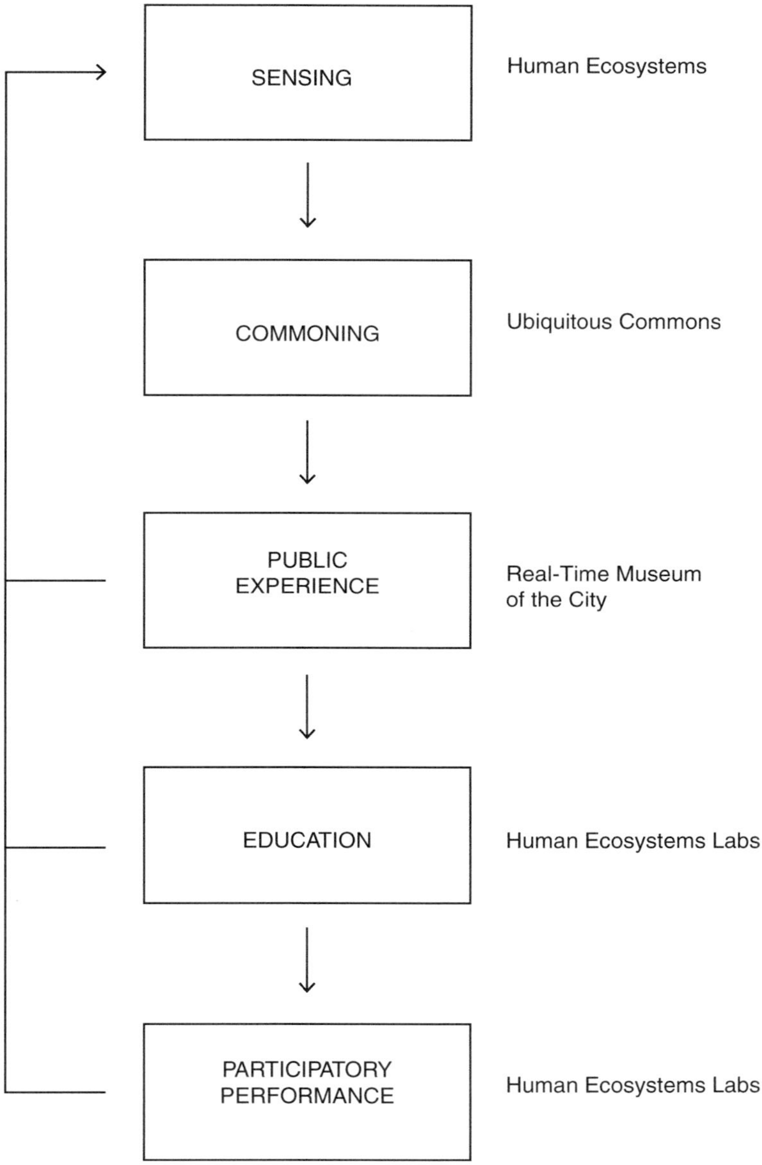

Fig. 9.1 Digital Urban Acupuncture methodology diagram

usually comes under the form of forms, databases, surveys, census, statistics, or even leaks of documentation. Then, there is data originating from technical and technological infrastructures, for example, energy, mobility and communication, which usually comes under the form of the streams of data captured by sensors, the

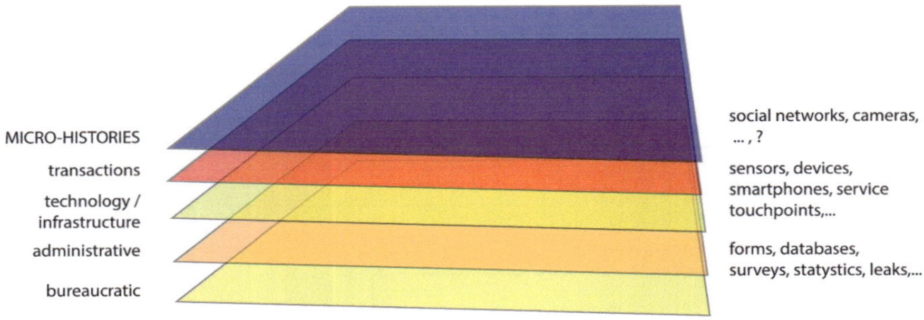

Fig. 9.2 The layers of information in the city

data collected at service access points, geographic information systems, and more. Next comes the data and information created when using transactional systems, such as for credit cards, toll booths, access systems, devices, smartphones, wearables. Last there are the data and information, which are generated through the microhistories of cities, as explained in previous chapters, and which have a wide variety of ways in which they can be sensed, for example, through social networks, cameras, movement and presence sensing systems, physical interactions, sounds, and more.

Moving from one layer to another, it appears clear how these forms of data and information progressively transform from being extremely explicit, formal, encoded, and consciously produced, to completely implicit, informal, and produced unconsciously. If when filling out a form at some office of the city's bureaucratic system, we may be able to explicitly recognize the fact that we are generating data/information, and this notion becomes blurred when we are placing a call with our mobile phone (after all, we perceive that we're just making a call, not generating data about our location, quality of service, duration of the call, its destination, etc.), or completely opaque when interacting with other people on social networks or, even more, by traversing a public square or going shopping while being continuously filmed by dozens of CCTV cameras, from multiple points of view.

On top of that a large part of the data, information, and knowledge we exchange does not have a direct digital counterpart. Chatting at a bar; overhearing a conversation on a bus and getting an idea from it; exchanging random remarks with random people in the street; watching people go by; sitting in the sun on a public bench: all of these activities can potentially generate enormous amounts of precious information about the ways in which people use the spaces of the city and in which they exchange data, information, knowledge, and emotions, but they do not have any direct digital counterpart which can be captured, analyzed, and shared. They are unforeseen, emergent, and serendipitous. But, if we were able to record them, they would be able to draw the patterns and to describe the relations which city dwellers generate in their daily lives, inhabiting the city.

These issues convey the need to adopt different modalities to collect the information about the Relational Ecosystem of the city. While some of these modalities can be quite straightforward (although not simple, mainly for the need to make sure that the various sources of data and information are actually compatible with one another, that they make sense when used together, in the same information environment), like accessing a database or a transactional system's log, others are not, and need specific tools, practices, and approaches to be collected.

Two families of approaches can be described for the more emergent, informal, and serendipitous forms of data/information generation: artistic/creative and inclusive/engaging.

On the one hand, artistic and creative approaches can be used to subvert usage of spaces and times of the city, in order to establish alternative ways in which people are able to generate data and information and which can induce them to do so through fun, desire, aesthetics, suggestion, suspension, and other mechanisms. For example, arts projects like Candy Chang's "Before I Die,"[5] or our "Consciousness of Streams,"[6] or the many more which have provided people from all over the world with entertaining, fun, suggestive, beautiful ways in which to express their emotions, sensations, opinions, desires, and actions in cities all over the world are precious examples of the ways in which arts and creativity are able to create shared states of expression and performance which can generate valuable data.

On the other hand, more direct—even manual—and inclusive approaches can be used to enable people to express in the city.

Figure 9.3 shows a manual version of the Human Ecosystems framework used in occasion of a workshop held in 2014 in Castrignano de'Greci in Apulia, Italy.[7] This system is "manual," meaning that the data it harvests does not derive from social networks, sensors, or other digital devices or means, but it is generated directly from the people who interact with it, by drawing on the screens installed in urban space. We used this technique because of the particular characteristics of the city and of its population. Castrignano, in fact, is a town with several peculiarities: It is small; people are abandoning it; there is a high level of digital illiteracy; it hosts several forms of immigration, from political refugees coming from the middle east and northern Africa to immigrant laborers; and last, it is one of the last places on earth in which Grecanico is spoken, a language which is a direct derivation of ancient Greek. Trying to observe the city through its digital expressions (for example, through social networks or sensors) would have made little to no sense: Only a minimal part of the local population would have been able to express and, thus, to show up on the emergent observation. For this reason, we adopted a different modality. We designed an interview format and a simple toolkit which

---

[5]Chang C "Before I Die" http://beforeidie.cc/site/about/.

[6]Iaconesi S, Persico O, Art is Open Source, "CoS – Consciousness of Streams," http://www.artisopensource.net/network/artisopensource/projects/cos-consciousness-of-streams/.

[7]For more information: http://www.artisopensource.net/network/artisopensource/2014/07/30/communication-knowledge-and-information-in-the-human-ecosystem-p2p-ethnography/.

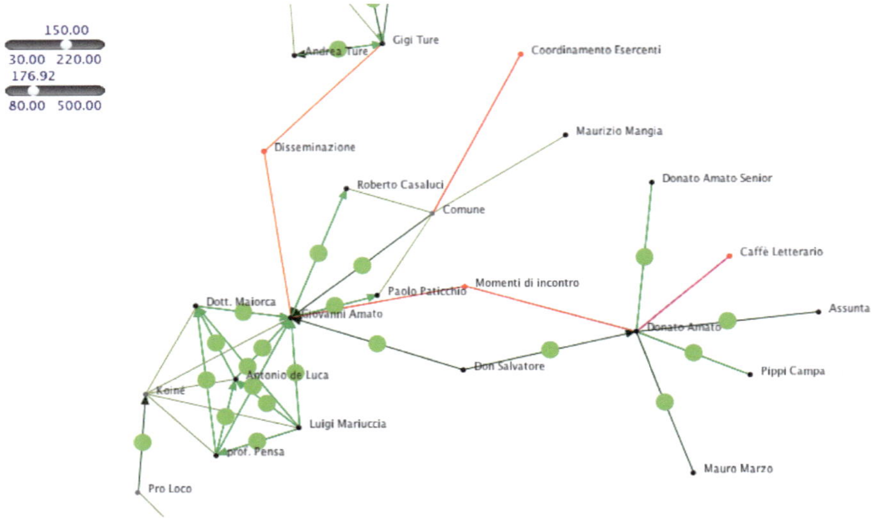

Fig. 9.3 A manual version of the Human Ecosystems, used in 2014 at Castrignano de'Greci, in Apulia, Italy

featured screens which were disseminated in the urban space and on which people could draw the Relational Ecosystem of the city. We then went around the city, establishing relations with the local population, collecting interviews, and teaching them how to draw or contribute to the drawing of the Relational Ecosystem, by drawing the nodes which represent people and organizations (like the Pro Loco visible in the image, which is an organization caring for local tourism and promotion of the territory), and lines connecting nodes to express relations. This manual modality proved to be very effective: On the one hand, interviews allowed to establish relations with people; on the other hand, the simple, accessible digital tools allowed to augment the literacy in the population and also enabled the emergence of autonomous expression and social–relational mapping processes. The process led to an exhibit and to the generation of multiple ideas, including the creation of a literary café.

The second phase of the DUA method deals with *commoning*. The data/information/knowledge captured from and through the Relational Ecosystem, as well as the process for capturing it, are transformed into a commons. This means that not only the harvested data, but also the process to harvest it are made available, accessible, and usable—even through replication, by enabling citizens to start their own capture processes—to all the stakeholders. In this phase, it is of major importance to avoid focusing only on the resources, the data, and information, but also include in the picture the high-quality relational environment which is needed in order for these resources to be meaningfully and effectively governed by the collectivity. As described in earlier chapters, a commons is the unison of these two elements: the resource (the common resource pool, using commons' terminology) and the high-quality relational environment. They go together and the first is useless without the second.

In some cases, it may be sufficient to release the data under the form of Open Data, using one of the many available open standards and ensuring their accessibility.

On the other hand, most of the times it is not. Administrations releasing Open Data have systematic problems in dealing with the inclusive characteristics of this practice: They tendentially focus on the technological and technical apparatus which is needed to release such data, and on the adherence to the many standards and quality indexes provided by the many ranking systems promoted by the international open government community. The intervention at cultural level and at the one of imagination is usually minimal: Efforts are dedicated to publishing the data, but not to making the data—and its implications—desirable, usable, meaningful, a driver for the imagination of wide areas of the population. This causes the majority of people to become spectators of this process, waiting for someone to eventually do something with this data—maybe some information visualization on some newspaper or Web site— in this scenario Open Data becomes a novel television, and people the spectators of the data show.

On top of that, most of the data released under Open Data schemes can be virtually meaningless for the majority of citizens. Yes, it could be interesting to know that my city administration has spent more on hospitals than it has spent on street lighting, but the average citizen will have no clue about just how much money constitutes an adequate expense for hospitals, and how it relates to the one for lighting. The result is that a mediated experience is needed, and experts may have very different opinions, may interpret data in very different ways, they may have very different political agendas, they may represent very different factions, and so on. The result is like television: Someone else takes information and extracts their own interpretation or meaning out of it, packages, and represents it for you, the spectator of the show.

Furthermore, the data and information which is more meaningful for people, the one which is able to set the social, anthropological, and cultural context for the living scenario is completely omitted. There is plenty of data about which percentage of the annual budget the city administration has dedicated to this or that. There is no information at all about if Mrs. Jones is worried about her job, or if Michael is happy about his school, or if Mary had angry because she has no place to play grass hockey, or if Mr. Rose is feeling generally anxious, all the time, for no apparent or understandable reason. All these missing information could be mapped, charted, visualized, counted, composed with other data sources, discussed, made the object of a consultation or of a shared decision making, in ways which are much more meaningful to citizens, and in ways which could enable better understandings of the city, made by its citizens. Where is all of this information? It is in the microhistories of the myriads of citizens, the big absence in Open Data.

For this phase, in our model—and as described in previous chapters—we use the Ubiquitous Commons, which enables citizens, administrators, professionals, organizations, artists, designers, activists, and more to associate the data they produce to their expression about how they'd wish this data to be used by specifying one or more identities—whether individual, collective, anonymous, temporary, nomadic, or a combination of these— describing both the scope and meaning of the data/information (e.g., "this data is for civic use") and the relations described by it (e.g., "this data is for the fellow citizens of my neighborhood"), at the same time enabling citizens to assume responsibility, to express their imagination and desires, and to build social, cultural, and human capital.

For all of these reasons, the Relational Ecosystem of the city—composed by administrative, bureaucratic, infrastructure, transactional and microhistoric data, information and knowledge, in its constant evolution—is transformed into an immaterial commons, ready to be hosted in the high-quality relational environment of the city.

How is this high-quality relational environment created? Through the following steps of the method: *public experience*, *education,* and *shared performance*.

Starting with the *public experience*, which directly answers the question: How can we transform the Relational Ecosystem of the city into an object/process which is desirable, beautiful, and relevant for wide areas of the population?

The answer is simple: By forming a novel sense of aesthetics—of beauty, using a slightly less adequate word—and, thus, of possibility—of what a city is and can be—and by including this as a fundamental, iconic, suggestive part of the public space.

In our scenario, this step coincides with the Real-Time Museum of the City, as described in the previous chapters. The Relational Ecosystem becomes a museum, a public space which is accessible, comfortable, and immersive, and where the Relational Ecosystem becomes a tangible, visible experience. People can see it represented under the form of info-aesthetic visualizations and installations using sounds, objects, generative sculptures, and more.

Figure 9.4 shows the Real-Time Museum of the City in Sao Paulo. The museum is a public space, free to access and to use. Citizens, tourists, schools, and all other city dwellers can come there to experience the real-time city and also inspect its past through the visualizations and installations. The museum is a place of imagination, which opens up the mental spaces through beauty and aesthetics. Not only this, it represents also the possibility to reclaim the digital public space of the city, as here the data, information, and knowledge contained in the infosphere of the city become a public resource, leaving the walled gardens of the operators just as they leave the perceptive distance of the Open Data portals provided by the administrations. The information is there, visible through the visualizations, and experienceable as a suggestive, iconic part of the public space.

In the museum, people can learn to ask questions to the city combining topics, places, emotions, and languages: "Who is afraid for their job in this neighborhood?"; "Who loves his school and speaks spanish?"; "Where do people who speak Chinese go to have fun?"; "What communities are expressing anger for public transportation issues?"; "What do people mention when they express fear for their safety?"; and more.

This possibility leads us to the fourth step in the method: education.

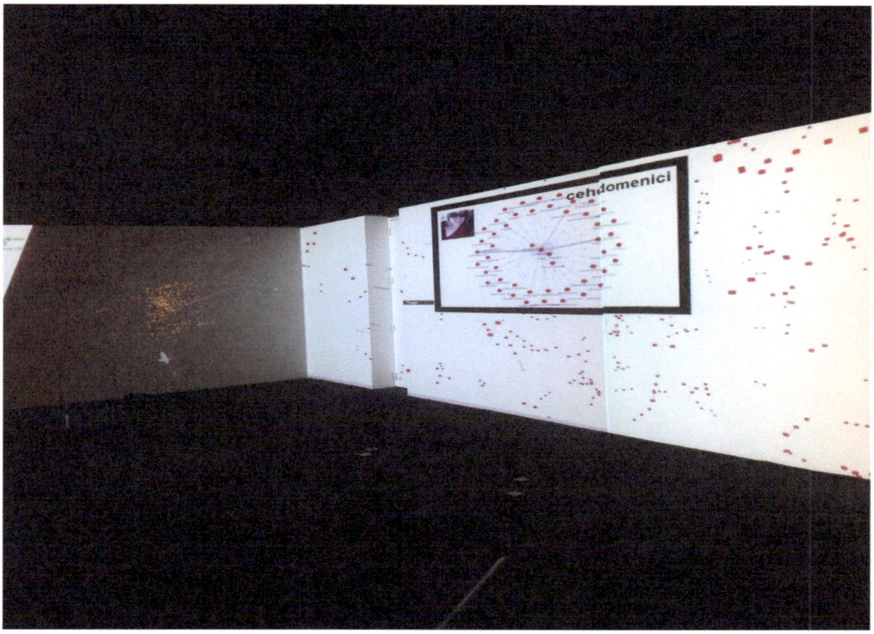

Fig. 9.4 The Real-Time Museum of the City in Sao Paulo, Brazil

In this public space, the Real-Time Museum of the City, all sorts of people find opportunities to take part in *education* processes. Children, elderly, students, researchers, public administrators, managers, artists, designers, planners, and more: Everyone learns to use the Relational Ecosystem of the city.

This is a very important step, as it leverages the mental space created by the aesthetics and experiences of the museum—the transformed perception of what the city is, represented as a constantly changing relational network between people, spaces, buildings, nature, infrastructures, times, events…—and uses it to start a constructive process: People are given the tools and knowledge to operate in and for the Relational Ecosystem, to achieve their desires and expectations, and to open up participatory processes.

In Fig. 9.5, we can see one of the workshops taking place in Sao Paulo. There, in the museum, completely immersed and surrounded by the many manifestations of the Relational Ecosystem and of its flows and transformations, across all layers of data, information, knowledge, and emotion in the city, the educational experiences take a variety of forms. Children learning how to understand more about their city. Students making a collaborative school project. Artists creating generative artworks which use data and information to create sculptures, sounds, visuals, all reacting to people's behavior and expression in the city. Designers conceptualizing toys, services, applications, from emotional compasses which allow to emotionally navigate the city to applications which promote specific forms of tourism or deal with

Fig. 9.5 Workshops in the museum, in Sao Paulo

emergency scenarios. Citizens understanding how to create shared decision-making processes and how to start peer-to-peer actions in the city. Public administrators learning how to start participative policy-shaping processes. Researchers doing research in linguistics, anthropology, sociology, and statistics. Planners understanding how people use the spaces in the city, as well as their desires, expectations, fears, anxieties, and joys. Entrepreneurs learning how to build new services which are not only financially sustainable, but which are also able to create positive social impact in the city. Elderly people learning to discover the digital life from which they are often excluded, and discovering ways in which to do things together with people, to interweave generations, and to generate shared value from their knowledge and experience.

All of these and more examples aim at one purpose: Leverage the beauty and aesthetics of the public presence of the Digital Urban Acupuncture process to inspire people's desire and expectations and to feed it with the tools and techniques which they can use to start doing things together. To take data, information, knowledge, and their flows out of the laboratory and into society: Eliminate the figure of the "scientist outside the system," looking at a petri dish, and inject a new version, in which sciences, humanities, planning, and governance are processes which happen throughout society, in participatory, shared, and ecosystemic ways. The scientist and the expert assume new roles, entering society, and transforming themselves in catalyzers, in shamans, Casagrande would say, who conduct a ritual dance of knowledge, opening up the doors of perception for people, to be able to recognize a different world, composed by flows and opportunities to intervene on them in collaborative ways, to create change.

Digital Urban Acupuncture is an enabling method, opening up mental spaces and enabling people to use it to create relational interventions in the city, with institutions and governments acting as catalyzers, enzymes, and to use Clément's phrase, "gardeners" of a "moving garden," a "garden without a form," just like the continuously evolving human and natural landscape of the city.

The last phase is the continuous catharsis and completion of the previous ones: the *participatory performance*.

As described in the last few paragraphs, the DUA process aims to enable the emergence of the perception of the Relational Ecosystem of the city to augment its quality and meaningfulness, and to use this renewed high-quality Relational Ecosystem to collectively manage a peculiar immaterial commons: the Relational Ecosystem itself. All of this enable citizens—with the catalytic support of their institutions and of other possible partners, playing the role of the "gardeners"—to perform the city through emergent, spontaneous, numerous micro-interventions. The performance of the city, in participated ways.

The "needles" of this type of acupuncture are to be found at multiple levels:

- in the "connecting the dots" action, in which the understanding of the Relational Ecosystem and of its flows allows to understand how to connect, interweave, and entangle individuals, communities, temporary or topic-based social groups, emergent social formations, with their roles, games, languages, imaginations, expectations; the needles, here, are the spots in which to interfere, mediate, connect, intercede, along the complex network of relations, to create new flows;
- and in the interventions themselves, in the participatory performances, in which all of the stakeholders have learned how to use the Relational Ecosystem and its flows to make things happen in the city from artistic performances, to activist processes, to urban design interventions, to policy-shaping processes, to shared decision making, to new service creation and deployment, to collaborative action, to fun, entertainment, company, and more; here the needles are in the spaces, times, contexts, topics, emotions identified through the transformation in the Relational Ecosystem, in its flows and mutations, and on the insights which we can gather from them, by learning how to ask questions to the city and obtaining a time-based networked answer in return, with myriads of points of view expressed, tensions, connections, separations, harmonies, dissonances, synchronicities, a temporalities, and how to interpret these networked answers in collaborative, connective ways in order to stage a participatory performative action.

All of the phases change the system.

The public experience stage mutates the perception of the city, and thus its Relational Ecosystem. The education stage transforms people's knowledge and, thus, the ways in which they perceive the public experience stage and also the

Relational Ecosystem. The participatory performances radically transform the Relational Ecosystem and thus, the public experience and the education stage, which benefit from mutated cultures, languages, imaginaries and even tools and services and techniques which can be among the outcomes of the performances themselves. All of these things transform the commoning process, and the commons itself, as it derives from the sensing stage of the method.

A second-order cybernetic system forms in this scenario, with multiple levels of nested feedbacks loop, representing the transformation of the city.

9.2 Forces in the Ecosystem

DUA revolves around the possibility to observe the Relational Ecosystem of the city, and to be able to understand its flows of communication, information, knowledge, over time, across space and being able to recognize roles, clusters, bridges, and modalities within it.

As we have seen in the previous sections, we can define roles according to the positioning and behaviors of nodes in the ecosystem. These classifications allow to identify specific types of subjects by analyzing simple behaviors over time and content. These are connective behaviors that describe the ways in which these subjects relate to others according to the ways in which they produce, exchange, curate, or aggregate information and knowledge. It is the description of a peer-to-peer learning environment.

How do communities and cultures emerge from this type of analysis?

It is a complex, unstable process. Cultures and communities change all the time, and they are not univocal in their description. Boundaries are fuzzy and blurry, and constantly change, as people discover the world and create relationships, forget about them, find affinities, establish conflicts, discover new hobbies and interests, develop passions and hatred, and more.

To gain a better understanding of how things work, we can imagine the space of relations as a physical space, complete with its forces.

We can imagine “forces of gravity,” which react to the various characteristics of the subjects in the Relational Ecosystem.

They are both “forces of attraction,” but work according to different logics.

The first is a similitude gravity, and the second one is the coexistence gravity.

The similitude gravity reacts on how “things are the same.”

If we consider the type of data which can be collected (as shown, for example, in Chap. 4), we will be able to see how a force of this kind could work for the two main types of data: the quantitative ones (and the quantifiable and categorizable ones) and the networked ones. For the first ones, a concept of direct similarity would apply (how close are two numbers; if a content belongs to the same categories…). For the second ones, a concept of networked similarity would apply (how a network of nodes of different types interconnected through links of different weights are similar to one another, for which there are a number of different algorithms present in technical literature; in general, in this second intendment, two nodes attract if the relational graphs with which they are connected are similar, for both their form and the characteristics of the nodes; degree of attraction is directly proportional to this kind of similarity).

The coexistence gravity reacts on how "things can live together without conflict."

Using this logic, the two types of information (quantitative and networked) will behave according to different schemes.

Quantities and categories will behave according to the logics of complementarity: small will attract large and vice versa; A will attract B if B implies or complements A and vice versa, etcetera.

Networks (topics/emotions and relations) will behave according to the logics of compatibility: Two people will attract each other if they express on a certain topic with similar or complementary emotions; and they will attract one another also if their relational networks are not in conflict with each other.

This consideration is to be handled with special care. At a certain moment in time and in a certain context, when can two relational networks be said to be not conflicting? This is a concept which is very different from similarity, as it does not imply "sameness," but "compatibility."

Figure 9.6 shows the logics behind the two types of forces.

Two networks do not conflict when they have complementary or non-overlapping architectures (nodes and their links), and the characteristics of their nodes are complementary or independent.

Compatible, non-conflictual networks fit like pieces of a puzzle, or are parts of two different puzzles.

Here, of course, the idea of "conflict" is a potentially misleading one.

A person, for example, could desire conflict (literally): For example, user A on a social network may express the desire to argue about a certain soccer team, while another might be a fan of that team. This would actually make the two profiles compatible, none conflicting.

Another typical example using social networks is the "troll" (a user who specifically enjoys conflictual engagement with other users): A troll could be easily classified as a hub for their capability of selecting their victims, or even as an expert, for the high relevance which people attribute to his responses (failing to adapt to the famous "do not feed the troll" rule). Sometimes trolls may be even

Fig. 9.6 Forces in the Relational Ecosystem

perceived as bridges, for their ability to interconnect and create engagement, even across communities, in the attempt to stop them.

As we have seen through this simple example, these same logics can be conveniently reversed to understand conflicts within Relational Ecosystems. For example, by understanding non-similarity, it is possible to discover factions. By understanding non-compatibility, it is possible to discover isolation. By observing both and, thus, their repulsion (e.g., if similarity attracts, non-similarity repulses), it is possible to observe conflicts as they evolve. By understanding their linear combinations, it is possible, in general, to discover a wide variety of conflictual or aggregative social behaviors. At the same time, it is possible to observe their dynamics to understand how to make interventions in these forces, to mitigate or extremize their effects, to allow for more inclusiveness, tolerance, culture clashes, Diasporas, and more, providing richness instead of stress, flow instead of stasis.

All of these considerations, of course, do not only refer to social networks, but also to sensors, devices, organizations, and non-human entities. For example, a seismic sensor, as we have seen in the previous chapters, can be sometimes described as an expert and, in this role, may be part of relational networks which may be similar or compatible with other ones.

These activities and flow can be observed and also acted upon in a digital version of Urban Acupuncture, in which the pressure points in the communication and information flows in the city are searched for in the digital public space defined by social networks, devices, and other ubiquitous technologies.

It is, of course, a mistake to fall to the temptation of describing these Relational Ecosystems as purely deterministic systems.

As we have been able to observe in our analyses, not all behaviors are explicable using these simple laws and classifications. Continuing with the metaphor of the physical universe, the universe of relations also has other forces which act at different scales and with different sensibilities, and also indeterministic forces. Other forces we might want to define on purpose to highlight specific behaviors (just like it is possible to define new roles, as seen in the previous chapters).

These forces are usable, useful approximations of the complexity which happens in the “real” world. And as such they should be used.

We could even focus on the possibility of a plurality of “real” worlds, just as it happens with quantum physics and relativity. In this case, the other worlds would not only be different reference systems, but also different definitions of time, space, context, and society. It could prove to be a useful exercise, one to do on purpose to easily confront with specific issues, just like changing a reference system may greatly simplify finding the solution to an equation. What happens when we observe these ecosystems by changing reference systems for time, for example, taking 1 ms, or 1 s, or 1 day, or 1 year, or 100 years, or 10,000 years as a unit? Different scales would pop up in all of these cases, allowing to focus onto the instantaneous or onto the geological, or somewhere in between. The same would happen if we would change the reference for space or topic.

Just like in physics, we could even start exploring the Quantum Physics of Relations, or maybe a Theory of Relativity applied to Anthropology and Sociology, including principles of Uncertainty.

While it might seem far off, it is a very interesting area for exploration, for example, starting from the Hawthorne effect, or from the Rashomon effect, or by the same observation of how ethnographers are not really "out of the system" which they observe and, thus, they contribute to the transformation of networks and relations.

And, in general, to direct research in an interesting direction, which is the one of inclusion: the elimination of the boundary distinguishing the observers from the observed, and suggesting an inclusive, participatory effort in the understanding of the dynamics of our ecosystem, to actively promote its well-being: a collaborative performance of coexistence.

From our point of view, this direction of research leads to a possible definition of what we call *P2P Ethnography* which, as ethnography, can be defined as a qualitative research design aimed at exploring cultural phenomena.

Different from ethnography, its aim is not to produce field studies or case reports, but to establish continuously available, diffused, accessible, participatory, performative, and collaborative processes which allow gaining understandings about the knowledge and the systems of meanings in the lives of a social group, and its interactions with other ones.

This is done by creating a commons, to which anyone can participate, in a performative, participatory observation of the cultural phenomenon, and in which all of the observations are accessible to everyone, manageable, remixable, and recombinant.

P2P Ethnography represents a participatory, performative approach, in which research and understanding require gaining awareness of one's position within the Relational Ecosystem (from cultural, emotional, aesthetic, perceptive, cognitive points of view) of the observed social group, and to establish or modify relations and interconnections both within the group, outside of it, and in-between, in fluid, dynamic, possibilistic ways.

In the next chapter, we will see the Digital Urban Acupuncture method in action as well as an application of P2P Ethnography.

9.3 An Interview with Marco Casagrande

Marco Casagrande is a Finnish architect, environmental artist, architectural theorist, writer and professor of architecture. Marco Casagrande is the laureate of the European Prize for Architecture (2013), CICA Award of the International Committee of Architectural Critics for conceptual and artistic architecture and UNESCO & Locus Foundation's Global Award for Sustainable Architecture 2015. He is the head of the Ruin Institute. He is well known for his definition of the Third Generation City and on his practice in Urban Acupuncture.

Ruins constitute syncretic maps of urban environments, as they show the citizens' usage patterns, imagination of lack of it, traversals, and behaviors. In short, they constitute the map of the city as seen from the composition of the myriads of microhistories of its dwellers, and in its perpetual evolution and transformation. How important are these maps? What is their meaning and how can they be used?

These ruin maps are a pattern of Urban Acupuncture. They are small composts where the city is slowly fermenting. The surrounding city, at least the official city, may consider these composts as the smelly parts of the city and cannot really cope with them. The only solution which the official city seems to have is to erase them, turn the ruins into lawn or park, which was supposed to happen to Treasure Hill, for example. The official city is insensitive for the energy and potential which these areas are suggesting. Normal people though are highly sensitive for urban energies and would be fully capable to operate the map of ruins. The ruins are insulting the official control, but on the other hand they have the capacity of offering the most fertile topsoil for organic urban development, which of course seeks to get rid of the centralized power structure.

These ruins are voids in the mechanical tissue of the centrally governed city. They are openings to different times, values, dreams, and possibilities—like the attic of a house. It is very likely that the essence of the surrounding city would intensify within these voids and get mixed or at least in connection with other organic layers of the city, like in a black hole. These layers are invisible for the official control'. Ruins have the possibility of partly tuning the city toward the organic, toward the Third Generation City.

These ruins can be used in interpreting what the collective mind is transmitting. They can be sensitive platforms for local knowledge to emerge and evolve as receivers. They can be hot spots for new biourban knowledge building. One must be very careful with the relationship with the official city though. Once the political elements of the city will realize that something constructive, collectively touching and possibly media-sexy is cooking up in the formally smelly ruins, they will try to squeeze in and buy off the energy. In case the constructive energy of the ruins is sold, the construction will turn into destruction. The official city can only banalize the local knowledge.

When you say that the Third Generation City is the ruin of the industrial city, you point into a really interesting direction, indicating how the emergent, spontaneous, energetic dynamics of the city and of its inhabitants are useful in gaining better understandings about the city, and about the ways in which it is interesting to create interventions in its fabric. Maybe even more useful than the information which is obtained through administrative, bureaucratic, and commercial processes.

This also places citizens in a new light, suggesting ways in which their active participation becomes fundamental importance to understand the city, and to act in it.

How do you imagine citizenship?

How do you imagine institutions?

Every citizen is part of the big brain of the city. This collective conscious is complex, multi-layered, and organic, but it is still a sensitive nervous system. The official city wants to flatten these layers into a simple two-dimensional map of the city, which is the official reality. Citizens however move much more flexibly, freely, and multi-dimensionally in the city that is what the official map would allow. The official city is just a background for the real citizen activities. Unofficial information is powerful. Whole cities could be designed by rumors. Urban power structures want to flatten the multi-dimensional, resourceful, and somehow mystical real citizen. He is too much in connection with nature, and the industrial city wants to claim independence from nature. Nature is seen as something hostile, something that wants to break the machine, and the untamed natural citizen is an unpredictable agent of nature, an urban native that needs to get civilized, needs to be saved from himself.

Institutions should be the inner organs of the city to keep life pumping through it. They could also be partly the nervous nods, which are dealing with the information and other energy flows of the city, thinking of the collective mind. City is one brain. Also nature is only one brain. City should be part of the natural one brain and the institutions should take care of that. Now the institutions are human-focused in a controlling sense and separating the city from nature. The solution which the developed institutional city is offering to the citizens is mechanical and standardized life. City should be a biological man-made organism and part of nature, otherwise it is against nature, a mental disorder—a human error as one might say. Institutions should be organic and most likely modular. They should treat the urban organism through punctual interventions, which would then be connected with the nervous network of the urban brain. This cannot be based on control and hardness—those are death's companions. City is not an institution, it is a living organism. Accident is greater than human control.

Relation, conversation and flows: Your vision of the city is very focused on these themes. Even in unexpected ways, for example, using the term of "urban rumors." This is very interesting, as it does not imply a concept of beauty and value which is not centered on form, but, rather, on the presence of energy, dynamics, fluidness, and emergence, and also on the harmonies and dissonances, the conflicts and consensus which are typical of living ecosystems.

How can an urban planner, a public administrator, an architect, a designer, or a citizen learn to recognize this new aesthetic and this value, and use them for collaboration, participation, action, and performance in the city, with other people, institutions, and organizations?

How does this fit in with your connection of Urban Acupuncture?

A rather good example of this was the cooperation between the Ruin Academy, JUT Developers, and the Taipei City Government. First of all, the Ruin Academy was set up as an open platform for different universities, disciplines, and professionals to participate in multi-disciplinary research and design workshops, courses, and actions. The topics for these assignments would be developed together with the City Government and in the end we would also report to the City Government. Still again, this was all unofficial. The City Government would not officially commission us and we would not be tied to any official nor academic bureaucracy. We would have access to the official data and intelligence, but we could operate much lighter and direct more like the Special Forces. Our operations were financed and backed up by the JUT Developers and the participating universities, who would also benefit of our findings and developments.

All the participating universities, the Tamkang University Department of Architecture, National Taiwan University departments of Sociology and Anthropology, and Aalto University SGT Sustainable Global Technologies research center found it very fruitful to have an open academic operational platform, which did not belong to any university, but was more based on academic squatting. Our interface to the surrounding city was also more real than with the locked academic disciplines, and the interaction with local knowledge proved to be vital to the new knowledge building of the Third Generation City.

City governments are full of departments and disciplines and every corner has a king. These kings do not talk to each other, but still again they are the first ones to admit that the highly regulated and protected administrative hierarchies are not optimal for mostly cross-disciplinary and multi-layered challenges and possibilities of the urban reality. They used the Ruin Academy to say things that they cannot say and to study things that they cannot study. We could have meetings in the evenings with the city officials, who could pass us the questions, interests, or notions on which they wanted the Ruin Academy to react—almost as if they would be operating with a clandestine organization. We would be their interface to the unofficial and to the underground, to the normal.

In some cases, like with Treasure Hill, the City Government was using me as kind of a joker, wild card. In Treasure Hill, the Park Department of the City Government was already destroying the unofficial settlement, when the Cultural Department of the same City Government was commissioning me to save Treasure Hill. In the end, they were all evaluating, how did I succeed and decided to come smiling with the results. "This is exactly what we were commissioning you to do." If I had failed, they would have just blamed the stupid foreigner and bulldozed down the place.

What I am trying to say is that we need to develop more unconventional ways to deal with the urban problematics. We already have the city governments, administrations, organization, NGOs, and universities, which are taking care of the official routines, but we need more flexible, straightforward, and light operators to work avant-garde and behind the lines, also underground. Operators who can make hearts and minds connections with the local knowledge and mobilize people to develop their city. Operators who can communicate with the shared mind, the collective urban conscious.

Urban Acupuncture is both a strategy for urban development through punctual interventions and straightforward tactics. There are many holes and cracks in a city and these cracks can be used for cooking up the operations. In the end, the city of cracks is much more interesting and humane than the two-dimensional, flat, industrial-modern city. Urban Acupuncture is breaking the industrial control, but it is for good—it is constructive anarchy.

What transformation comes into this scenario when many people in cities have integrated digital means in expressing their microhistories, relations, emotions, behaviors, in conscious of unconscious ways?

Commercial intelligence has been using the methodology of Urban Acupuncture for a long time. A good example is the network of 7-Eleven convenience stores in Taipei. They are located in carefully studied commercial acupuncture points around the city, building up the densest network of 7-Elevens anywhere in the world. Our actions, behaviors, wishes, desires, and individual histories are constantly monitored, traced, and processed based on our digital activities. Our digital networks have the power of launching rumors and revolutions, but they are also very easily manipulated. An interesting question is, what is the interface and dialog between this digital mind and nature? Can it support the organic knowledge as a portal for the collective mind to communicate? Are we now just looking at the simplified and flat prints of the digitally moving information the same way as the official city is flattening the informational space of the city into a two-dimensional map? Would it be needed to penetrate through the thin layer of visual information surface to the actual digital space, where the countless informational layers are generating new streams of knowledge and how can we communicate with this digital subconscious? I think that this subconscious wants to surface on the city. It wants to take both form and be sensed through our physical presence and our natural mind.

In your conception, the Third Generation City is itself a form of knowledge. This is yet another parallel with the concepts expressed in Digital Urban Acupuncture, where the Relational Ecosystem of the city, captured through data, information, and knowledge exchanges, becomes a commons, available and accessible for everyone to use.

How, in your opinion, should this knowledge be accessible and usable?

And, on top of that, how is it possible to suggest and create the basis for the emergence of the imagination, sensibility, and desire to use this knowledge?

This knowledge is already now a source of intelligence for political and economic power speculations. It is also a form of new culture. Physically, millions of people are now migrating based on this data—migrants, refugees, and people moving to cities. Big digitally formed and manipulated tribes and armies are in physical war and one of the main frontlines is digital.

The digital realm is one surface of individual and mass communication, but is it yet a form of new knowledge? The digital underground movements and paths seem somewhat hopeful, but the big data is just entertainment and commercially controlled—not very different from the official city. One should not be blindfolded by the online access to information and entertainment. Flesh is More.

How important (or not) is education in your vision of the city? How can the literacy and sensibility which are needed to conceive the possibility of accessing the knowledge in the Third Generation City emerge in citizens? Through a school? An educational process? Peer-to-peer processes? The availability and accessibility of tools and methods? How?

The unofficial community gardens and urban farms of Taipei are run by anarchist grandmothers. Also the urban farming communities are often matriarchal—

like Treasure Hill. The ex-Soviet collective farms are by now run by babushkas. Modern city is a patriarchal structure as a form of industrialism. In Taipei, the kids go to help on the collective farms—carry water, dig soil, etc. Sometimes they come to the farms after school to do their homework. They learn how to farm and the local knowledge becomes real for them. One step away is the official city. Modern man should take the liberty to travel a thousand years back in order to realize that the things are the same. What is real cannot be speculated. What is real is valuable. There is no other reality than nature.

Maybe the digital realm is also nature. Possibly it is like resonating behind the singing of the birds. We can either listen to the birds' singing or we can feel and contemplate with the resonating behind it. Maybe we are resonating with the digital flows as well. We can feel the mind, but cannot really interpret it. Nature is a life-providing system. My friend is a digital monk and he is very much tuned with nature as well.

Third generation citizens do not need to be educated. They already exist. They are the ones connected with local knowledge and sensitive enough to feel the different pulses and messages of the city. They are the mediators between the city and nature. We all have that quality, but we are educated to forget it. The third generation condition requires us to forget the forgetting.

Instead of education, we need to learn how to pay attention to the seeds of the Third Generation City. We need to document them, learn from them, and let them grow. Most of all we have to stop ignoring them. These seeds are often cooking at the unofficial layers of the city. City is a big compost, which needs to be turned around every now and then in order to keep it alive. The seeds of the Third Generation City are in connection with the local knowledge and they form a pattern of organic Urban Acupuncture to the static city trying to tune the urban development into biourbanism. The challenge is that these seeds do not necessarily support the economic speculations and can be quite contradictory to the established power hierarchies, which try to suffocate them. Hence their existence in the underground. The digital realm may be a possibility for the local knowledge, for the seeds of the Third Generation City, to communicate and connect with the larger mind of individual citizens.

How? Forget the forgetting. Industrialism is young and simple. Let the organic growth make a new layer on top of the industrial city. This coexistence will develop into the next step of urban industrialism, and the city can learn to become an organic machine. In a sense, we must ruin the mechanical city and open up the industrial control, so that nature can step in, nature including human nature.

References

Casagrande M (2008) Cross-over architecture and the Third Generation City. Epifanio 9. http://www.epifanio.eu/nr9/eng/cross-over.html. Accessed 31 Oct 2013

Casagrande M (2010) Urban acupuncture. http://thirdgenerationcity.pbworks.com/f/urban%20acupuncture.pdf. Accessed 31 Oct 2013

Casagrande M (2013) Third generation City. http://casagrandetext.blogspot.it/2013/10/third-generation-city.html. Accessed 12 Nov 2015

Iaconesi S, Persico O (2014) Urban acupuncture in the era of ubiquitous media. J Community Inf 10(3)

Jacobs J (1961) Death and life of great American cities. Vintage Books, New York

Lerner J (2014) Urban acupuncture. Island Press, Washington, DC

Lydon M, Bartman D, Woudstra R, Khawarzad A (2011) Tactical urbanism: short-term action long-term change, vol 1. The Street Plans Collaborative, New York

Chapter 10
A Case Study: Rome's Cultural Ecosystems

Abstract The Human Ecosystems and Digital Urban Acupuncture approaches have been used by Rome's Public Administration to gain better understandings about the city's cultural ecosystem, to enact public engagement directed to promoting participatory decision-making and policy-shaping processes, to foster the emergence of peer-to-peer self-organization patterns, and to create novel forms of economies. The project began in 2013 by observing the Relational Ecosystem of the city for 6 months, then opening up the information (which is still being captured and used to this day), configuring it as an Open Data source, and creating an open laboratory accessible to citizens, administrations, organizations, associations, companies, and multiple types of cultural operators, which could learn how to use the Relational Ecosystem to achieve their goals. This chapter describes this experience, its results, and the directions in which it is currently heading.

The dataset for the Rome case study is available at http://www.human-ecosystems.com/DAP/DigitalUrbanAcupuncture-CaseStudy-Rome-dataset.zip.

On September 2013, Rome's city administration started the ECM(1), "Ecosistema Cultura Municipio 1" (Cultural Ecosystem of the 1st Municipality) project using Human Ecosystems and the concept of Digital Urban Acupuncture.

The objective of the project was to start a participatory process across the city which would involve as many cultural expressions as possible (cinema, music, art, theater, dance, events, presentations, independent circuits, new media, architecture, etc.) and as many types of people as possible, from operators, to artists, designers, architects, citizens, PRs, and everyone.

This was the first official instance of the Human Ecosystems project and technological framework. The objective was to create a shared human, relational, and organizational geography of the city which would account for as many cultural expressions, communities, cultures, languages, and space/time contexts in the city.

The process was to follow the DUA methodology: First, create a sensing framework, to be able to listen to the city in all desired modalities; second, turn the results of the observation and of the listening into a continuously updating immaterial commons; third, add this commons to the public space creating a public

S. Iaconesi and O. Persico, *Digital Urban Acupuncture*
DOI 10.1007/978-3-319-43403-2_10

presence for it, through visualizations, events, and public meetings; fourth, teach people how to use it through education processes; and fifth, enable people to come together and make things happen, through shared, participatory actions.

The objective was to implement a framework for the collaborative development of the cultural ecosystem of the city.

10.1 The Sensing Framework

The manifest need was to observe the cultural ecosystem of the city, where "culture," in the intentions of the municipality, indicated the cultural activities of the city.

This is a very wide definition. What is culture, in this sense? Is sitting on a bench in the park culture? Is cooking culture? Is going to church culture? Is a hackerspace culture? Is a cinema culture? And so on. Everything people do, in a sense, can be defined as cultural activity.

We decided to focus to adapt to what the administration expected.

We focused on these themes in the city:

- arts;
- music;
- performing arts (theater, dance, etc.);
- public events and entertainment;
- languages;
- education, edutainment, and infotainment events;
- public presentations;
- new media and technology activities and events (including the immaterial ones);
- architecture;
- design;
- literature and poetry;
- countercultural activities (independent/alternative circuits, DIY productions, etc.);
- protests and manifestations;
- events in public space.

For each of these, we designed ways in which we could observe, tendentially, all of the actors (whether individuals, professionals, organizations, operators, institutions, or else) in their public expressions in the city of Rome.

We designed:

- a way to preload the initial list of actors and to observe their public expressions and communications, as well as their impacts and effects;
- a way to extend the initial list by focusing on emergent and spontaneous expression and communication;

- a way to include internal/external influences, much in the same way shown in Chap. 7 for the Kansas City case;
- a way to sense new topics and interests, in order to further expand the observation in interesting directions.

The initial list was composed in by combining multiple sources: the registry for cultural operators in the city as maintained by the administration; a thorough search on major social networks (Facebook, Twitter, and Instagram) for evidence of professional operators in the aforementioned topics and for people attending their events; a Web search on the topics and locations; a list of known locations in the city where cultural events of the listed kinds regularly take place (to be used as geographically named entities in the natural language analysis components of Human Ecosystems); and a list of sources—such as online magazines and revues—which advertise cultural events in cities (with the objective to use them as sources to sense the ones happening in Rome).

The lists were cleaned up (e.g., removing duplicates or dubious and unreliable results) and merged.

Table 10.1 shows the numerosities of these categories (in the table, the results are disjoint, meaning that if a certain entity appears in a row, it does not appear in the following ones).

The elements of the initial list were used as the starting point for data and information collection using Human Ecosystems. The social networking profiles of all of the subjects who possessed them were configured as the initial set of users in the Relational Ecosystem. 4 % of the subjects in the registry had no social networking presence: They have been set aside to be added in the next phases using manual techniques and were added to the natural language processing engine as named entities, to be able to detect mentions.

At this point, the expansion mechanism was put in place, to be able to sense the spontaneous and emergent expression of the city. For this, two mechanisms have been put in place.

The first manual allowed people and organizations to add themselves to the observed user base, using an online form, by self-declaring their role as cultural operator or user in the city. The online form was advertised and the entries started coming in, totaling 15 new operators (professionals and organizations) and 120 new

Table 10.1 Numbers of subjects observed in the initial phase of ECM1

Type	Quantity
Registry of cultural operators	713
Professional operators on social networks	1482
People attending the events as demonstrated from social Network posts (2-months lookback on social networks)	46,396
Locations for cultural events	594
Sources for event advertising and reviews	117

individuals. The entries were verified before being added to the databases through an online check: Many were already in the database (240 operators and 536 individuals); many turned out to be spam, bots, and other unsuitable identities (7499, also following a spam wave which hit the online form).

The second one was automatic. Human Ecosystems was used to configure and run a wide harvesting process for people using social networks from the city of Rome to mention events, shows, presentations, lectures, many other forms of cultural events which were relevant for our observation, or the locations in which these events were held (through the geographically named entities). Natural language analysis was used to analyze their content to understand whether they were organizing them, attending, participating as artists, the times, dates, and modalities of their participation (Were they going alone or together with someone? Where did they come from? and more). Social networking interactions and natural language analysis allowed to reconstruct the relationships between all of these users, as seen in the previous chapters. This allowed us to store classifications about them and to verify and update them as more content was harvested. The harvesting process was run continuously, and by the end of 2014, it had identified 962 new cultural operators and 266,194 cultural users.

The same technique was used to capture the external relations of Rome's cultural ecosystem, composed by the people and organizations who were talking about Rome's cultural events, were coming to Rome for an event (or were passerby while they attended), promoted an event, took part in one (where do the artists who play/show/exhibit in Rome come from?), and more.

A similar mechanism was also used to collect all of the online resources which advertise cultural events, to be used to discover new operators, venues, locations, and events which were both used directly and in the natural language analysis as named entities.

All of these techniques were also fused together to provide the detection functionalities to detect new and emerging topics, which were placed in a separate classification in respect of the ones agreed with the city administration, ready to be analyzed separately, and, if of interest, added to the initial list.

Figure 10.1 shows the geographical distribution of the data harvested in this way, consisting of more than 3 million social networking messages generated by more than 80 thousand social network users and revolving around 15 thousand events across more than 800 locations.

As the image shows, maximum concentrations take place in specific parts of the city, especially in the historical center of the city and in a few other locations, mainly in the San Lorenzo, Pigneto, Prenestino, Trastevere, Testaccio, Saint Peter/Castel Sant'Angelo, Prati, Parioli, and Garbatella neighborhoods, which are the ones in which the principal concentrations of events and venues are.

A different scenario is shown when the map is made to highlight the various types of contribution, differentiating the content generated from operators, users, and mentions of venues, as shown in Fig. 10.2.

Fig. 10.1 The geographical distribution of the harvested data, at the end of December 31, 2014

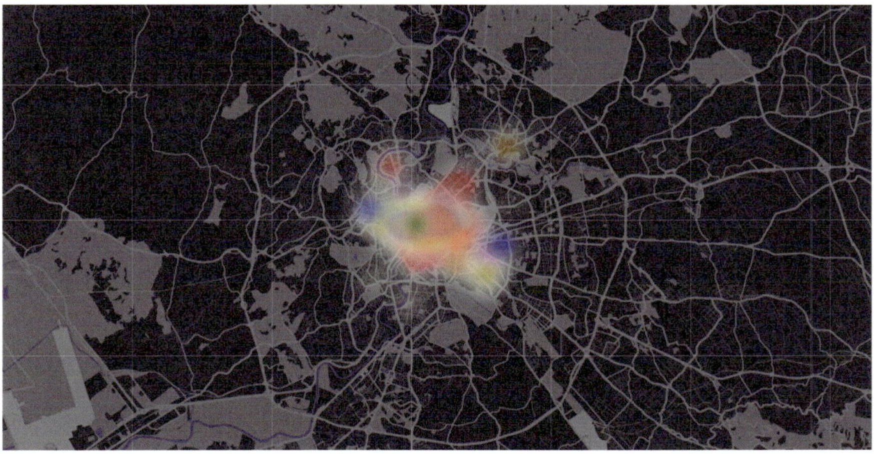

Fig. 10.2 The geographical distribution of data, according to multiple classifications

In this image, for example, red shows the maximum concentrations of operators headquarters (the places from where they emit their social media content), blue venues for literature, yellow for music, orange for theater, green for history- and art-related events, and so on, with many different classifications being represented on the map. Additional visualizations can be built to show different distributions.

As the data harvesting process was prepared with its automation layers to expand the reach and scope of the capturing as well as its fine-tuning to ensure the maximum relevance for the proposed objectives, the analysis processes were put in place.

The objectives were interpreted to define the following requirements for the result of the analysis:

- *Relations*: Highlight the Relational Ecosystem of "culture" in Rome, showing the relations among operators, professionals, institutions, and citizens; their roles; the flows of communication, information, knowledge, and opinion; and the evolution through time;
- *Time*: Highlight the patterns in time; the times when activities/events of the different kinds concentrate; the peaks and lacks of events of different types; and the rates of attendance;
- *Space*: What are the destinations in the city for topic/genre X? What are the origins (where people come from) for topic/genre X? How do people get to events? How do people collaborate to get to the events? Are there limits and barriers to get to the events? Who/what stimulates large and small crowds to move across the city?
- *Communities and Markets*: How do people form recognizable groups which are definable as communities or markets? Do they overlap? Is there permeability and/or mobility from one group to the other? How inclusive and/or closed are the various communities? How does communication, engagement, and information propagation diffuse among and across these groups?
- *Remix*: Enable to mix all of the previous perspectives among themselves and with other sources of data, enabling answering questions like which are the audiences for a certain type of action? Where do people for event X come from? What is the return on investment for event X? And for operator Y? Are different operators collaborating? Are they "stealing" audiences from each other? What/who influences communities? What is the shape of "alternative" markets and how do they relate/interact with the commercial/institutional ones? What do people desire? What is on their agenda? How is it possible to communicate topic X? How is it possible to create community engagement on topic X?

These were just some of the questions which we came up with, and additional ones would come in the following stages.

For each one of these questions, a data structure to contain both its answers and their evolution in time was created.

Some of the most fundamental ones (meaning that they formed the foundation of the whole observation and the base for the other ones) were already contained in the Human Ecosystems infrastructure (e.g., everything referring to the Relational Ecosystem).

Other ones had to be designed on purpose.

For example, the question is "what are the destinations/origins for event/action type X?" This is a very important question to answer, as it can have impacts on public transportation; distributions of events in the city; facilitated funding for events of a certain type in specific locations with high demand and low offering; or the opposite, if we feel that a certain type of event is underrepresented in a certain area of the city, and we want to promote it. All of these and other impacts show the relevance of this question for both policy-makers and citizens (who might, for example, use the result of the "what communities arise around topic/genre X?" to enact a collaborative crowdfunding process to bring a certain type of event in a specific location, possibly with the support of the city administration using equity funding or some other method).

For this question, a simple data structure was designed, listing all of the events and their types/genres—as recognized from the sources and from the parsing and natural language analysis of the content harvested from social networks—and a relation table connecting them to the coordinates of the venues in which they were held—recognized through geographically named entities—to the ones of the places people were coming from to attend them—recognized by combinations of various geolocation techniques and, in general, by using statistical distributions to guess people's "home" location according to the places where they most predictably generated their content. Each record of the data was characterized by a time stamp, in order to be able to generate time series and to recognize patterns in time, as well. The simplified data structure used to collect the answers for this question is shown in Fig. 10.3.

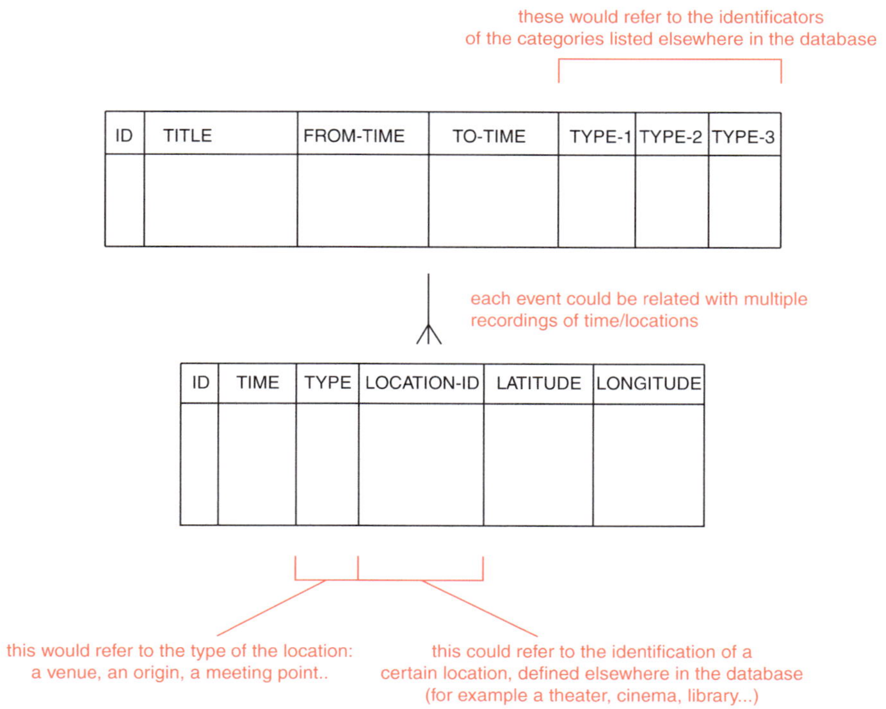

Fig. 10.3 The simplified data structure for the question "what are the destinations/origins for event/action type X?" question

Another important question is dealt, for example, with the social impacts brought on by a certain event or cultural operator. This could be translated into the analysis of the variations in the Relational Ecosystem in regard to an event (or to all the events executed by a certain operator). It could be algorithmically determined in the following way:

1. take a snapshot of the Relational Ecosystem before the event (meaning store the current nodes and relations, the perceived roles of the nodes, and the current intensities of the flows of information, knowledge, emotion, and opinion);
2. take a snapshot of the Relational Ecosystem immediately after the event;
3. start from the nodes in the Relational Ecosystem who expressed something about the event (whether they attended, liked or disliked it, expressed that they wished they would have been able to attend, etc.) measure:
   a. the new or transforming relations appearing in respect of the previous snapshot(s) with other nodes who expressed about the event and their characteristics (persistence, verticality, horizontality, transversality, liminality, weight, etc.);
   b. the transformations in these nodes' topics and topic/emotion networks;
   c. the propagation patterns for information, knowledge, communication, and emotion for all the topics relevant to the event;
4. take a snapshot of the Relational Ecosystem 15 days after the event;
5. repeat Step 3 for this new snapshot, comparing it to the previous ones;
6. take a snapshot of the Relational Ecosystem 3 months after the event;
7. repeat Step 3 for this new snapshot, comparing it to the previous ones;
8. in general, periodically repeat snapshots and analysis for the required time.

By analyzing and visualizing the results, important insights can be drawn about the cultural impact of a certain event. The events from a single cultural operator, in a certain neighborhood, for a certain theme, in a certain timeframe can also be combined to achieve broader types of analysis.

For example, we performed this type of analysis on the TechCrunch event held on September 26 and 27 at the MAXXI Museum in Rome: an international event about technological innovation and disruptive business models.

The results of the analysis—performed through a sequence of 4 snapshots to the Relational Ecosystem taken immediately after the event and 15 days, 3 months, and 6 months after it—are highlighted in Fig. 10.4 which shows the variations compared to the initial snapshot, taken before the event.

The snapshot taken immediately after the event shows that 2542 nodes (and, thus, people, organizations, or other subjects) had radically transformed their relations, the things they talked about, and the ways in which they exchanged information, knowledge, and emotions. This is quite normal: They had just been through an event, messaging to each other, retweeting, liking, linking, connecting, and networking. Many were disconnected (the isolated nodes), meaning that they changed what they were saying, but not explicitly connecting with others.

Fig. 10.4 The visualization of analysis of the impacts of the TechCrunch event in Rome at the MAXXI Museum

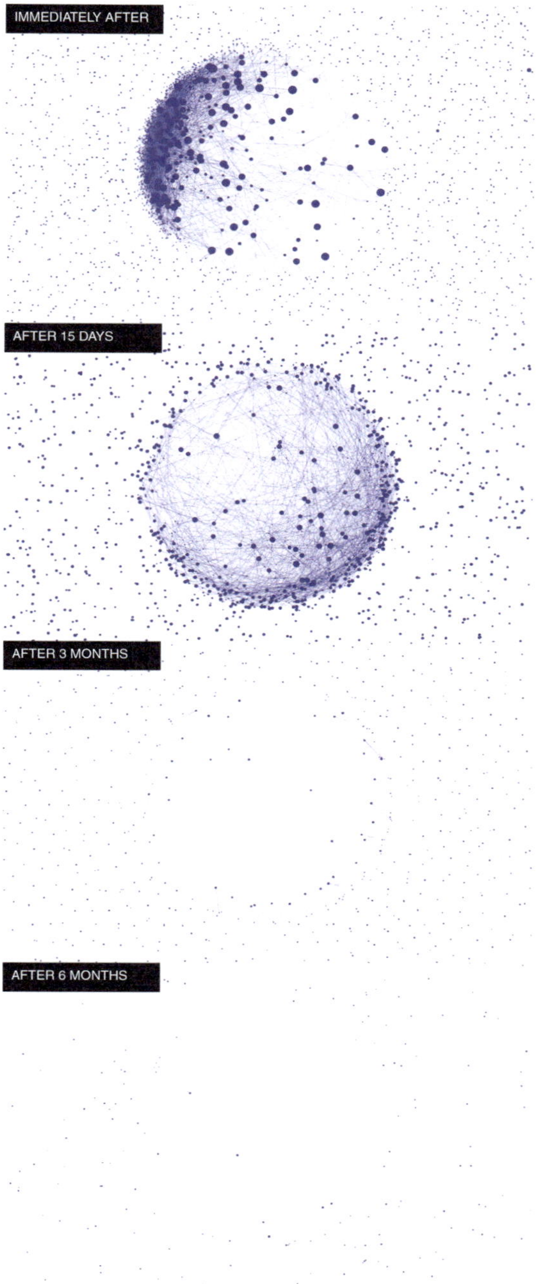

Many were very polarized in their intensive connections: They formed large, intense clusters in which everything was shared, forwarded, and commented multiple times (as visible on the center-left side of the image, where the connections are numerous and intense).

After 15 days, two weeks after the event, the situation was very much alive: 1412 nodes maintained the transformation (in respect of the situation before the event). The clusters of connections were somewhat less intense, and there were still many of the disconnected nodes: People who were not in relation before the event were still actively chatting and exchanging information, and many of the people who evidently transformed their topics of interest during the event maintained the transformation.

After 3 months, the situation was radically different. Only 612 nodes remained transformed. The connections were few and far apart, and only a fraction of the transformed among the isolated nodes kept on expressing on the topics they had adopted during the event.

This trend was confirmed after 6 months, when only 164 nodes were still showing signs of the transformation, with very few (almost none) relations still showed and a very sparse crowd still showed signs of the cultural transformation.

What does this mean?

It means that the event had a somewhat small cultural and relational impact on the city. The people who attended were mostly related among themselves before the event and that a very limited number of new persistent relations formed, apart from the ones immediately after the event. It also means that the event provided a limited amount of transformation in people's discussions and topics of interest: The people who attended and who were not already dealing with these issues before the event were excited at first, but quickly (3 months) slowed down (almost stopping) expressing about these topics.

It does not mean that this was a "bad" event. It means that it created a limited number of new and persistent relations in the city and that it changed the perspective on the world of a very limited number of people. Most of the people who attended the event were already interested in the issue confronted by the show and mostly already knew each other, and they were part of the same tribes: All the rest were mostly passersby, leaving practically no trace. The cultural impact on the city—if we define in terms of the transformative effects on the population—was very limited.

These are just two of the analyses which were produced in answer to the many questions in which we formalized the objective for ECM(1), which show how we can use the Relational Ecosystem to confront with issues. In all, they were 52, applicable to single cases or parametric according to time, geography, topics, emotions, subjects, languages, and roles.

We are now ready to focus on the next steps of the DUA methodology, as it has been applied for ECM(1).

10.2 Commoning

The next phase was to transform the Relational Ecosystem and the data, researches, analyses, information, and knowledge it produced into an immaterial commons.

In this case, this was done in a very straightforward, minimalistic way. A repository was created to host the data and all of the other artifacts as they became manifest: documents, visualizations, and data answering questions as highlighted in the last part of the previous section. It was available until 2 months after the end of the project, being shut down at the beginning of March 2015.

The materials and data were made available according to three logics: a document and multimedia repository; an Open Data repository; and a set of APIs (Application Programming Interfaces) for real-time querying of the data.

The document and multimedia repository contained all of the documentation, visualizations, reports, spreadsheets, tables, and Web-based interactive experiences which were produced during the project. It was a minimalistic structure of online folders organized by item type and date, with the possibility to search for keyword, dates, topics, item, or filetype. It was accessible for all citizens without restrictions or need for authentication, and the content was distributed under a "Creative Commons, Non-Commercial, Attribution v.4.0" license.

The Open Data repository was organized much in the same way, with folders hosting periodic exports from the databases as well as of the data structures relative to the 50 questions described in the previous section of this chapter, organized by topic and date, and with the possibility to search by keyword, dates, and topics. The licensing scheme was the same as before.

The APIs were documented on a series of Web pages. They consisted of a series of RESTful method invocations which produced results in JSON format. They were completely accessible without the need for authentication, but had frequency-of-invocation-based limitations, to make sure that the systems would not suffer overloads and that they would not incur in denial–of-service attacks. A maximum cap on the amounts of results that were obtainable through a single Internet address was also in place for the same reasons.

10.3 Public Presence

The project did not have the budget to implement the Real-Time Museum of the City. For this, a set of alternative strategies have been to be implemented to materialize the public presence of the Relational Ecosystem in public space, to suggest the emergence of novel imaginaries and visions for all stakeholders, and to foster the formation of a high-quality Relational Environment in which to operate.

Different patterns of action were used:

- temporary exhibits;
- public stakeholder meetings;
- media and communication;
- integration with other events, happenings, and contexts.

Temporary exhibits have been created for the Relational Ecosystem, in order to create suggestive interventions in the urban fabric which could alter the perception of the life of the city for citizens, tourists, and other city dwellers.

With the visualizations available online, it was relatively simple to organize public showcases by using the many spaces which are available to the city administration, to set them up using projectors, monitors, and sounds, and to simply turn on a network-connected computer using the online info-aesthetics at full screen. One of the visualizations is shown in Fig. 10.5.

These visualizations would be highly iconic, showing the territory of the city in a perpetual mutation according to the transformations in the city's Relational Ecosystem: The map of the city would transform its shape; create mountains and depressions where specific emotions would be expressed; assume different colors where certain topics were discussed or represented; become connected across its territory with arches of lights as soon as relations would become manifest; and emit sounds as the emotional landscape of the city would transform. They would mutate according to various formats and representations, shifting from maps, to graphs, timelines, matrices, and more. These were very important interventions, because the experience of seeing the city represented in these ways turned out to be very communicative with people, providing them with insights about all of the ways in

Fig. 10.5 One of the visualizations used in the temporary exhibits

which it could be reinterpreted, opening up their mental space about what constitutes a city and how it can be operated.

The temporary exhibits were also useful tools for communication and relation, from three points of view:

- first, to be able to have a cognitive hook to communicate the upcoming stakeholder meetings; when experiencing the visualizations we would, in fact, let them know about the dates and times of these meetings, either by speaking or by always having information materials close by for them to take; this proved to be an effective solution;
- second, to communicate that it was the people themselves, the citizens, including the one whom we would be talking to, who were generating the visualized forms of the city, as we were seeing them right now; the opportunity to influence a representation of the city as shown by a public institution proved to be an empowering feeling for citizens; when they realized that they could alter what the screens and projections were showing by emitting a single tweet or message on Facebook, most people felt empowered and immediately started expressing ideas on the ways in which they could use them to complain, to suggest approaches, to express needs and desires;
- and, third, because people could actually find themselves in the visualizations; this was an important feature, both to create the evidence of the fact that they were the ones driving the visualizations through their expressions, and for the fact that this was an instantly effective way of confronting with the problematic issues connected with these kinds of practices; when they asked "what am I doing there, exhibited on the visualization?" we would answer "come to the meeting and we will both explain how to protect your privacy and how to use all of this information and data for positive reasons, to be a better citizens, to collaborate with other citizens, and to make things happen in the city."

The stakeholder meetings (e.g., the ones shown in Figs. 10.6 and 10.7) were public gatherings in which the administration would invite all sorts of cultural players in the city, from operators and professionals to regular citizens with an interest in the cultural activity of the city. They would be held in theaters or in other public spaces which were usable by the administration, to be able to organize them on a low (almost non-existent) budget in order to allow for spending on materials, documentation, activities, small exhibits, and performances, to create a welcoming, joyful, rich environment for mutual confrontation. The Human Ecosystems was on show, too, to create a continuity with its other manifestations in the city.

The meetings would start with the presence of relevant city councilors and representatives from the local government, which would introduce the day. The events would last 6–8 h and would be divided into a discussion session, in the morning, and into an operative and educational session, in the afternoon.

Fig. 10.6 The public stakeholders' meetings

Fig. 10.7 Another stakeholders' meeting, in a theater

The discussion session was started by the public administrators, and it would quickly turn to the presentation of the findings obtained through the Human Ecosystems and its Relational Ecosystem, always including in the discussion the pointers on how to obtain the data, what to do with it, and how to access education on this topic and more.

The discussion would then open up: The citizens could speak freely either starting from some of the observations coming from the Relational Ecosystem, or introducing their own topics of interest. While this was happening, lists of relevant issues were composed by collecting them from the observations of the citizens, and they were also researched immediately using simple tools connected live to the Relational Ecosystem, in order to be able to expand them in the discussion by introducing the opinions and expressions of the rest of the city.

This was a really interesting feedback loop, as it managed to create suggestive dynamics in which the room in which the gathering was happening was transformed into an open window to the city and also because people consistently found the expressions of the "rest of the city" as interesting and relevant—whether they agreed with them or not—and able to raise the quality of the discussion. This fact also constituted a passage ritual toward the education experiences: People wanted to know more.

One more way in which the Relational Ecosystem entered the public sphere was through media and communication. Articles on newspapers, initiatives like the OneMillionDreams project,[1] wide discussions on social networks, and talks and presentations inside museums[2] and libraries[3] and in other locations and contexts all became media objects and environments which managed to infiltrate the process into public sphere and into people's consciousness. The opportunities and possibilities entered citizens' perceptions and, thus, assumed form into the public sphere: Human Ecosystems, and the way in which it had been used for ECM(1), was a recognizable process which could open up the perception of the city.

Furthermore, multiple collaborations were established with other initiatives, integrating the DUA process in other contexts, connecting it with other communities, and finding/generating new spaces for reflection and collaborative action: from exhibits (for example the one shown in Fig. 10.8) to lectures to university lessons integrated in multiple courses in different faculties, up to initiatives in social centers, squats, political gatherings, activist events, parties, art festivals and more. The Relational Ecosystem is integrated with multiple social groups to become a platform for the perception and understanding of the city.

---

[1]OneMillionDreams: http://www.artisopensource.net/network/artisopensource/2014/03/24/onemilliondreams/.

[2]http://www.artisopensource.net/network/artisopensource/2013/11/03/human-ecosystems-at-the-macro-museum-of-rome-for-aperitivi-formativi/.

[3]http://www.artisopensource.net/network/artisopensource/2014/02/17/human-ecosystems-at-artscaping-rome/.

Fig. 10.8 An exhibit of the Human Ecosystems at the Naked City festival in Rome

10.4 Education

The afternoon sessions of the stakeholders' gatherings were dedicated to the education phase of DUA. These were only a few of the occasions in which people could learn how to use the Relational Ecosystem, but understanding how they took place may be useful in understanding all of the other ones, as well. One of these education events is shown in Fig. 10.9.

There were two types of educational happenings.

The first type was dedicated to introduce people to the basic concepts of the process and dealt with the description of the harvested data and information, with the social/political/anthropological/legal implications deriving from the availability and accessibility of all of these data. People here were taught what data were being collected, how it was processed, what you could understand about the city, its relations and flows by using it, and also how to protect yourself from having your privacy harmed by these possibilities. Furthermore, people were being explained how to use these data, through ready-made visualizations and tools, by constructing their own, or by learning how to collaborate with other people.

These workshops and lessons, on the one hand, were dedicated to vision, imagination, and critique, introducing the main concepts, the implications, and their critical aspects. On the other hand, they were technical workshops, with people

Fig. 10.9 One of the stakeholders' workshop

learning how to interpret data, how to build tools and applications, and how to work together.

There were multiple versions of these workshops dedicated: to children, using simple tools and entertaining narratives; to students, showing some more advanced tools and instruments; to researchers, showing them development and analytical tools; to citizens and activists, showing them tools for collaboration and communication; to administrators and managers, showing them how to interpret the Relational Ecosystem to form strategies and to engage citizens; to artists, designers, and other creative profiles, to show them how to transform data and information into artworks, toys, applications, and services; and to elderly people, to show them how to understand this digital part of the life of their city and how to make sense of it and many more.

The workshops of the first type constituted elaborate introductions on how to use the Relational Ecosystem to "make something happen in the city."

The second type of educational process was dedicated to application to practical cases. Whether it was a citizen activation issue, something about public policies, something about a specific domain, an issue was defined and the workshop was dedicated to understanding the ways in which it could be tackled through Digital Urban Acupuncture.

These were practical, pragmatic processes: One issue was tackled, and Human Ecosystems was used to deal with it. It was an occasion to apply in practice what was learned through the first type of educational event.

Focus was dedicated to accessibility to these workshops. This was addressed using peer-to-peer-learning modalities. It was not necessary to have participated to one of the workshops of the first type, as there would be a number of people who did are ready to participate to the action and to take the burden of technical knowledge. At first, it was us and our collaborators, but as time went by, a progressively number of people had learned how to use the Relational Ecosystem and DUA, ready to support the others. While doing things, people learned, augmenting the reach of peer-learning modalities. New insights, tools, and visualizations were continuously being generated, adding to the shared toolset.

These workshops were peculiar in the fact that most of the people already knew which kinds of problems they wanted to address and were looking for ways in which to tackle them.

For example, one of the first stakeholders' meetings was dedicated to cultural operators. More than 500 of them attended. What happened was that these operators had very clear ideas about how to "fix" the cultural ecosystem of the city. These included, most of the time, the need to obtain public money for investments, shows, insurance for potential lack of audience, and more. It was never easy to shift from a model focused on subsidization to one centered on collaboration. Not everyone got it, but a few did. The ones who did were very passionate with it: When they understood the implications of the existence and accessibility of the Relational Ecosystem, they immediately saw opportunities for communication, marketing, audience management and engagement, competition management, collaboration with other operators and with citizens, and even arriving at the idea of possible crowdfunded theater seasons which could be driven through the Relational Ecosystem: No tickets sold; the money was put upfront by engaging citizens into the discussion about the structure of the theater season.

It was the incipit of the participatory performance phase of DUA.

10.5 Participatory Performances

As described in the previous chapters, the final phase of the DUA methodology is a performative one: Actors of all types can use the immaterial commons represented by the Relational Ecosystem of the city and use it, together with others, to design and implement a "needle," an intervention which function as a stimulation, as a reactivation of the flows in the city, a small shock which is able to establish an example which is able to change people's perception and to open up mental spaces, in order to cause wider effects.

Multiple types of interventions could and were made.

Some of these were more analytical and abstract, like the analysis of the TechCrunch event which was described at the beginning of the chapter and whose communication was accompanied by its corollary; the answer to the question is: "how do you design an event which is ecosystemic, meaning that it is able to engage wide communities and to enact wide cultural impact on the city?"

Other ones were much more pragmatic and practical (some are shown in Fig. 10.10 and referenced in the text below).

For example, in occasion of a particularly violent city riot, an application was designed by artists, designers, engineers, and political activists which would prove to be very useful in these dangerous situations. Using Augmented Reality, the mobile application allowed anyone to look around themselves using their smartphone as a lens, highlighting the directions in which episodes of violence were reported, in real time, on social networks. People caught in the middle of a riot could orient themselves immediately (green directions mean "no violence," and red ones mean "violence") and, thus, be facilitated in reaching safety, and they could also learn how to effectively express in the digital public sphere to help others in need.

These are only two of the examples, described with some detail because of their opposite characteristics: as cold as data and analytics and as hot as the fires of a violent citywide riot. But there were many more.

Some examples were: people trying to understand how to use the Relational Ecosystem to find other people to join them in cleaning up a park in their neighborhood; cultural operators trying to understand how to avoid stealing each other's audiences, and to find new business models; the first experiment in a crowdfunded theater season performed by the Teatro Valle; people finding ways in which to form complex collaboration partners between the arts, technologies, sciences, institutions and citizens; situations where the Relational Ecosystem was used in order to

Fig. 10.10 Augmented reality for dealing with riots in the city

promote the construction of collaborative decision-making and policy-shaping processes. And many more are creating virtuous feedback loops all over the city.

By the end of 2014, the Rome's city administration suffered from a series of difficulties and scandals and the project was dismantled. It remains, to this day, one of the first structured examples of the application of the Digital Urban Acupuncture methodology and one of the first examples of P2P Ethnography.

Chapter 11
More Than Humans

Abstract This chapter focuses on the ways in which non-human entities can be included in the Relational Ecosystem of the city, such as plants, trees, infrastructures, organizations, sensors, and devices. The methodologies, approaches, technologies, and modalities according to which these scenarios can take form are examined using a variety of approaches and examples. This chapter also points out the possibility to include different layers of ecosystems within the ecosystems, like in the case of organizations—for which their members appear in the ecosystem as both individuals and as the result of another ecosystem, the one of the organization—and in the case of distributed sensing networks—for example the natural environment of the city, composed through disseminated sensors across the natural spaces of the city. This chapter ends with an interview with Pier Luigi Capucci.

In the previous chapters, we have explored ways in which the human Relational Ecosystem of cities can be observed, analyzed, participated, and performed to bring positive change to the urban environment. In Chap. 4, we have also started noting how this ecosystem is populated not only by human beings and their communities, but also by a variety of other possible subjects, such as organizations, algorithms, sensors, the natural environment, and artificial intelligences. In this chapter, we will explore this possibility in further detail, understanding what are the implications of including these non-human subjects into the Relational Ecosystem.

In current times, it becomes progressively more difficult to establish precise boundaries between identities. Persons may represent their identities, because of their personality and the things they express, or because there is some legal and administrative form or document which certifies their existence, or because some database says so. We can have the experience of these people's identities by chatting with them on the street or in a bar, by viewing their posts on social networks, by observing their review of a restaurant in a certain city on a travel Web site, by remembering them, by seeing them on TV or on a newspaper, by hearing stories about them, and in many other ways.

Furthermore, using contemporary digital networked tools, it is fairly simple for a person to simultaneously adopt different kinds of identities, potentially appearing as

S. Iaconesi and O. Persico, *Digital Urban Acupuncture*
DOI 10.1007/978-3-319-43403-2_11

a multiplicity at our perception. Multiple social networking accounts, anonymity, and collective management of a single online identity are only a few possibilities among the many. Identities can be stolen, borrowed, and used to convey certain messages from some relational vantage points.

In general, human identities have become more complex in the digital era. They can, of course, be individual, meaning corresponding to a single body and mind. But, of course, a single body and mind can correspond to multiple identities. The simplest case in which this can happen, for example, is when a single person opens up multiple social networking accounts, with different profile data, relations, images, texts, and tastes. This can happen for multiple reasons: to express different interests; for a desire for privacy in certain contexts; as a joke on someone; as a fraud, trying to trick someone else into believing that a different person exists; for sexual reasons; for political and activist reasons, in which using their "administrative" identity could turn out to be dangerous or unwanted; to avoid employers to understand what their employees are doing online; and many more. In each of these cases, understanding what these persons are doing or achieving by building these multiplicities of identities is a complex thing. They are constructing, in more than one way, a different self, behaving differently, creating a "mask" to impersonate it, and establishing different sorts of identities. They may even choose to let some of their different identities to speak with each other, for example, to stage discussions, to have their arguments supported in some visible way, or as a joke. They would be wearing different identities as new skins, and these skins would also correspond to slightly (or enormously) different perceptions about the body and mind which is expressing them, both from the point of view of the self and from the point of view of how other people experience them.

In the same way, it is nowadays simple enough to achieve anonymity. By using relatively accessible and usable services and tools, it is possible, in fact, to assume anonymous identities. There are multiple toolkits dedicated to this possibility: temporary e-mail addresses which can be used to register to services; network anonymization tools; impersonation tools; network routing tools which allow messages to hop from one point of the Internet to the other before being delivered to the final recipient, in order to be untraceable; encryption tools; virtual private networks (VPN) services; and more. The result is that identity can become completely obscured in quite accessible ways. Many of the identities of which we have experience online are of this type. Most of the identities which are built in this fashion are completely invisible to us, as they are intended for specific usages and contexts. The fact remains that there are high numbers of them, for business reasons, political, and activist ones, for freedom of expression issues, or even just to have fun and to explore the Internet in unusual ways.

The possibility of multiplicity and anonymity opens up the doors to the possibility of collectivity. It is fairly simple to create and operate collective identities. Social networking services such as Facebook offer this possibility out of the box through their "Pages," "Public Figures," and more. But other forms are also possible. From the most simple ones like sharing the same user accounts among multiple people, to ones which are fairly complex, involving content distribution systems, discussion systems to decide what the collective identity should express and communicate, or artificial intelligences processing and assembling content coming from a variety of sources to fabricate identities with specific personality and attitudes which are, to all effects, the result of the combination of multiple ones.

11.1 And It Doesn't End Here

Identities can also be temporary, based on a specific context, event, scenario, or created for a limited time because of someone's desire or need. A temporary identity could be created for a festival, a fair, and a conference, to be able to concentrate all related communication in ways which can be recognizable, identifiable, and notable and without interfering with our everyday life identities of the different kinds.

Identities can also be nomadic, meaning that they can be given to someone else, transferred as easily as providing another person with the access credentials to some user account. This can happen for multiple reasons: as a business transaction in which, for example, a certain online identity has gathered so much momentum and influence that it represents an economic value and, as such, can be sold to a company to be able to use it to promote their business strategy; to handover a certain online presence, because the person(s) conducting it cannot anymore, and wishes the presence to continue; as a ritual; and even for death, when the online identities of someone who died are given to someone else to conduct them in their memory, or to continue their objectives, or to promote communication on certain themes.

All of these types of identity can also be combined to achieve multiple effects. For example, if an innovative conference would desire to offer its participants the possibility to express as a unison, it could create a collective, temporary identity which would be shared among the attendees. Or the mayor of a city could offer residents the possibility to participate to a collective, geographically limited identity, in order to create a shared sense of belonging and purpose, and to participate collaboratively to collective action. The same could happen at neighborhood level.

All of these types of identities can be stolen, borrowed, sneaked in, infiltrated, captured, replicated, and more, for a variety of reasons and causing multiple types of effects.

All of these possibilities, together, describe a very complex scenario which has effects beyond online domains, as people meet, consume, relate, come together or apart, collaborate, work, study, and entertain themselves in permeable ways, in and out of the net. An identity of these types has effects on our possibility to use a credit card, to buy a house, to get married, to have a job, to go to school, and to express emotions.

What this also describes is a major shift in our possibility to conceive our identities which was, of course, present in past societies, but which has boosted up with the advent of the digital society, in which multiple accessible, usable, and inclusive tools have broadened our perspectives, allowing us to do in a few clicks things which were practically impossible to achieve using only the physical realms of reality.

Ubiquitous technologies go even further in this direction, uniting the radical reconsideration of identity to a more general perception of space and time.

According to cultural anthropologist Massimo Canevacci,[1] digital culture challenges the classic distinction between space and time by promoting syncretism between both dimensions and by breaking with hegemonic dualistic thinking. His observation starts from the notion of ubiquitime, a neologism he created to define, from an ethnographic viewpoint, "the uncentered and nonlinear experiences of space–time" promoted by contemporary digital communication. In his view, the notion of ubiquitime combines three central notions: simultaneity, "an aesthetics made of fragments from metropolises and technologies," as advocated by futurism, the art movement; chronotope, established by philosopher Mikhail Bahktin to denote the dialogic relationship between the spatial and the temporal horizons that "has become essential to the development of literary polyphony"; and ubiquity, a metaphor that expresses the ability to be anywhere at the same time made possible by the potential for global connection of the global digital networks. Ubiquity is a key concept in digital culture, because it "characterizes human and non-human space–time relationships on the Internet." Ubiquitous individuals can move about between different identities, spaces, and times, giving rise to the multividual, according to Canevacci. The multividual emerges from the multiplication of subjectivities beyond fixed identities: "Ubiquity defies identity, which becomes more flexible. The ubiquitous subject of ethnographic experience is a multidividual."

The transformation of identity has wide impacts on cultural, psychological, cognitive, social, and political levels. It also effects on legal and administrative domains, at local, national, and international levels and at the ones which concern the Terms of Service agreements of the single service providers. Starting with the latter, the current opportunities for mutation of the concept of identity enter both the dimensions of conflict and opportunities for major service providers, such as Facebook. If, on the one hand, they match the possibility to more closely follow and understand user behaviors, to be able to transform them into economic value by offering dedicated services and through advanced data collection mechanisms, on the other hand they also represent a conflict in which users of these services become able to hide, shield, or obfuscate their behaviors, or to masquerade them, effectively detaching them from their "official" identities, potentially bringing down the value of the data which are collected and sold to marketers, statistics institutes, secret services, and more. Facebook's "Real Name Policy" is a controversial example of this fact. A wide controversy arose when the famous social networking provider started to enforce the policy according to which users were forced to use their "real"

[1]The following quotes are taken from the transcript of Massimo Canevacci conference at International Academia (ICA) Conference in Sao Paulo, on April 27 2015, available here http://www.iea.usp.br/en/midiateca/video/videos-2015/intercontinental-academia-talk-with-massimo-canevacci.

names on their profiles, the ones they have on their identity cards or passports. Ethnic minorities, LGBT users, established online alter egos, people using alternative names in their work, or social life have been affected by this phenomenon. While the Facebook's management expressed the desire to enforce such "real name" mechanisms to provide a safer environment and a better social experience to all of their users, it has affected people's freedom of expression, self-determination, and self-representation. Furthermore, it has opened a number of legal- and privacy-related questions, as well as a reflection about the shifts in power distribution which are represented by the fact that a multinational company is able to certify the identities of billions of people, requesting documents, proofs, and evidence, leveraging the high social pressure determined by complying or dying, at least at the level of sociality, by loosing one's account, and, thus, by relations, content, and part of the personal history. Reactions have been wide and in open conflict, so much that the social networking giant has announced several reforms of the policy.

To all accounts, several large, global players are actively pursuing policies, technologies, and mechanisms which deal with the possibility to cope with the transformation of identity in the digital era, both to transform it into an opportunity for business (e.g., the many services dedicated to anonymity, privacy, encryption, secure information leaking, and more) and to be able to maintain control and power.

As for the former element, instead, in regard to the legal and administrative issues which are being raised by the current mutation of identity, the most evident factor is also an obvious one: While transformation is happening—for example in the national and international legal systems, with the European Union leading the innovation for everything which concerns digital identity, privacy, anonymity, and other matters—laws and regulations change at speeds which are orders of magnitude slower than the ones at which societal and technological change takes place. This fact has powerful implications: from the fact that multiple innovations in service interfaces, practices, and technologies are implemented each year without the concrete possibility to understand them and their interpretations in terms of human rights and, thus, to devise proper laws and regulations for them—as soon as the law is in place, the innovation has already become obsolete—to the fact that a series of super-national organizations are, de facto, creating a new legal administrative domain which affect billions of people and which is self-determined by them, unilateral, and constantly changing with each microscopic algorithmic and interface transformation. Code is law, using Lawrence Lessig's expression. We have not yet understood the extent of this process, as society, governments, and human beings.

This fact becomes even more evident as new subjects enter the scene, with entirely new possibilities—and their corresponding dangers and opportunities.

In fact, digital—and, thus, physical—identities can become media for other types of actors to express themselves. There are a number of practical examples already in existence in which algorithms, software agents, and artificial intelligences—and the companies which express through them—can "use" our identities to act autonomously.

For example, "bots," which are those software programs which—either autonomously or by interacting with some existing online user account—automatically generate and publish content. Bots are used for advertising purposes, to send out notifications and even as spam messages, often flooding communication with generated messages which are intended to luring other users into making purchases or visiting certain Web sites or services. Bots connect, in authorized or unauthorized ways, with selected accounts and use the logics provided by their controlling software mechanism to pull out information from a variety of sources and use it to generate messages and notifications. Some bots are useful, for example, the ones dedicated to creating notifications about certain events or reminders. Other ones are damaging and even dangerous, as they can be the expression of illegal conduct or use users' online accounts in ways which are undesired. Both types have legal implications: If, for example, a bot would use a user account to administer illegal expressions, the user would be considered as guilty, and the burden of providing the evidence of his innocence would be placed on the user himself, essentially by proving how the bot usurped his identity in unauthorized ways. Even if the user was able to demonstrate this fact, being able to attribute guilt could be problematic, as bots of this kind usually operate from computer systems in foreign countries with permissive policies, and even the ownership of such computer systems could be obscured or unavailable. Who is responsible in this case? This is an open question and one to which legal systems do not provide certain answers for. To this day, if, for example, we were to be insulted or damaged by a bot and its expressions, it would be very hard (or impossible) to obtain assistance from our legal institutions.

Another interesting type of these software agents comes under the form of the many relation and community building services which multiple companies offer, such as Commun.it, Klout, Twitter Grader, TwitterCounter, Listorious, BlueNod, and the many more which offer such services. Users are able to connect their social networking profiles to these services to benefit from a complete set of analytics which allow to monitor online identities, presence, and topics discussions. Based on the insights produced through these analytics, most of these services are able to suggest relations to engage, discussions to participate in, conversations about specific topics, and so on. Even more, they are able to automatize these processes to an extent. For example, the Commun.it service is able to use the user's account on social media sites to automatically publish messages containing specific mentions of other users and of specific topics according to the logic of its algorithms, in order to optimize and expand the user's reach, community, and influence. The software agent, in this case, interacts and interferes algorithmically with the user's identity. It appropriates it, in a sense, abandoning the user's organic, natural, and spontaneous way of communication to adopt an algorithmic one (decided by the service), in order to achieve the service's objectives. This is an interesting leap: The software agent masquerades itself as the user, generating messages on his behalf, building the user's relational network. On top of that, the service includes itself in the mentions: Does it express the desire to "be friends" with everyone? Is it acting to ultimately build its own relational network? Do any of these questions make sense? To this day, the answers to these types of questions are open ended.

Using the same scheme, multiple other types of subjects, agents, and identities can appear and express themselves in the digital realm.

Using ubiquitous technologies such as sensors, cameras, networks, and actuators, basically anything can be connected to the network and communicate. Plants, buildings, appliances, furniture, toys, and other objects can be connected to a dedicated sensing system to capture some form of information from them, process it so that it becomes meaningful and comprehensible, communicate it over the network, and also to establish feedback loops in which other types of subjects become able to interact or respond to this communication—or to create initial stimuli—changing the object's state. People can now talk with plants, buildings, appliances, and objects, and they can answer back. All of these objects can also communicate among themselves, establishing relational networks of any kind: A plant could be equipped with a sensor measuring humidity and communicate this humidity to its owner and to a watering system which could turn on the water sprinkler; its owner could perceive the plant's "emotion" (a "happy" plant which is properly watered) and could influence the overall system by, for example, manually stopping the automatic watering, because he decided to water the plant itself and may be have a chat with it.

The possibility—once used in science fiction movies or in avantgarde artworks such as the ones by Eduardo Kac, Roy Ascott, John Lifton, Richard Lowengerg, Christa Sommerer, Laurent Mignonneau, and many more, or, more recently, by the Scenocosme and Data Garden groups, and the many more who, over the years, have adopted these kinds of practice—coming from the opportunity to transform plants and objects into sensible and communicative agents, which are able to translate their "feelings" and "expressions" in forms which are able to convey meaning to human beings and in order to allow human beings to continue the conversation and interaction, is an outstanding one. It allows to describe a scenario in which the Relational Ecosystem does not only include human beings, but also (tendentially) all of the natural and built environment, and the objects, devices, appliances it contains.

This progression can continue, to include agents which are—or progressively become—completely autonomous, controlled by algorithmic or "intelligent" logics. Diving into the discussion of what is "intelligence" is well beyond the scope of this book. What we can notice here is the ways in which perceivably autonomous agents and identities can establish relations with the rest of the Relational Ecosystem, which we have described thus far, and how they can actively participate to it, influencing their own behaviors and the ones of other members. We can start to investigate what happens when these agents enter "society."

It is very interesting to reflect on the fact that even if human beings have created algorithms and artificial intelligences (under the form of neural networks and other kinds of implementations), this does not mean that human beings are able to understand what these agents do (or think?).

A good, accessible example is Google's DeepDream image recognition algorithm. While using neural networks to recognize cute cats in Internet pictures, these algorithms search for recurring patterns and differences in ways which are incomprehensible for us. Not only we do not know what parameters, values, and contexts are used in this kind of evaluation (apart from what we know from the limited open source version which Google released): Even if we did know, they would make no sense to us, as human beings for which a 1 million record table is beyond understanding. They would not because these parameters are thought for a "digital brain," and they have nothing to do with how human beings think or operate. Even if they did make sense, we would not know how to call them and how to define the categories ("images with cats under the form of visual pattern number 1725342829"?) and the ideas behind the recognition of that certain pattern.

These are very important considerations, because they are the ones which determine how these non-human agents relate and interact with the others: When they recognize a certain pattern they act, if we are not able to understand what they recognize, and how they recognize it, and are not able to understand how they act when they do, it will be a problem to understand their roles in our Relational Ecosystem.

These misunderstandings also take place among these non-human agents. One example of this is represented by (Google's) Chrome Web browser and (Google's) YouTube services, and it is about automatic advertising. A sort of software war is going on between ad-blockers, online advertising, and the bots which fraud online advertising (e.g., producing false clicks to augment revenue). It is a very interesting war, fought with algorithms and emulating the behaviors of human beings. Within this billion dollar war, a series of interesting conflicts of interest are coming about, for example, the ones between the developers of the Chrome Web browser and YouTube's advertising, within the same company (Google, who draws its revenues from advertising). Chrome's developers are very passionate about the quality of their browser and on its security features. At the same time, YouTube developers have to live knowing that the developers in the next room, in the same company, developing Chrome, are developing the browser in ways which make the advertisements that they are developing useless. These kinds of tensions will get worse, and technical countermeasures will become more complex. A series of enterprises are already starting their own war, fighting ad-blockers.[2] In the meanwhile, publishers are starting a war on their users by inventing advanced techniques whose sole purpose is to show advertisements, no matter what.

The Internet is infested by all kinds of trackers, following users from one site to the other, cookies, and invisible pixels recording behaviors. Initially, the possibility to follow customers seemed as marketing heaven with employees not wasting half of their budgets on misplaced ads and having the availability of nicely quantified dashboards with all the insights they needed.

Then came the robots.

[2]For example, http://uk.businessinsider.com/anti-ad-blocker-startup-yavli-launches-2015-8.

In the beginning, they were simple: small pieces of code which loaded Web pages and automatically clicked the ads. Recognizing them was simple.

Soon enough robots learned to use real, remotely controlled browsers, making it more difficult to differentiate them from human beings. They learned to scroll and to move the mouse, just like human beings. Advertising networks started learning, then, how to contrast every robot innovation. They were learning how to detect the slightest signals to understand a human click. To do that, they designed intrusive software, invading people's browsers. Quickly enough robots began disguising themselves as people, just like the monster of the movies who wears someone else's face: Robots learned how to wear people's digital identities. They entered computers and use their Web navigation history, their cookies, and all of the tracking data living on those machines to perform as best as they could. Each iteration made them harder to recognize. They have become able to load dozens of advertisements onto a single pixel; to start Web browsers on hidden desktops with the sound turned off; to watch video after video, clicking on ad after ad; to visit real Web sites; to fill real shopping carts; and to stimulate the appearance of targeted advertisements which are more remunerative.

We now live in a Blade Runner like world in which robots of this kind pose as human beings and figures such as Deckard try to unveil them. To do that, they dig into our browsers, implementing javascript Voighy-Kampff machines to decide who is human.

We remain obviously caught up in this war.

Advertising frauds function because their market is highly automatized. It is very similar to algorithmic trading: Decisions are made in split seconds, and the correspondence between publicity and who sees it is beyond human control. It is also a complex world of lateral platforms to decide the offering and to understand demand; to retarget automatically; to pretarget; to model behaviors; to create real-time auctions for ad space; and to sell futures of advertising spaces. Payments are also automatic, and thus, it is very easy to come out of the game with cash: This is what robots do. Fake Web sites show real ads to fake users (or, better, robots) to earn real money.

Understanding the breadth of this phenomenon is difficult, also, because robot networks have learned to understand when they are being observed.

The winners of these wars are non-human, as well: large organizations such as Google, surveillance and control organizations, and the new ad-tech industry, for algorithmic advertising.

This long example is precious in understanding that when we deal with these non-human entities, we are not dealing with futuristic scenarios from the next century, but with the things of the present, embedded in our Web browsers, devices, and in the services we use everyday. These agents influence our behaviors and establish complex relations with us.

This will become even more problematic—and, as all problematic things go, also full of opportunities—when these intelligent devices will become more diffused in our societies, such as self-driving cars, domestic robots, smart refrigerators, autonomous biomedical devices, and all of the rest. The future of culture, psychology, responsibility, liability, crime, affection, and philosophy awaits us.

As of now, within the contexts of this book, we might want to explore how it is possible to include these non-human agents in Relational Ecosystems, whether they are plants, sensors, devices, databases, bots, artificial intelligences, and organizations.

Different methodologies, approaches, technologies, and modalities can be used in order to achieve this kind of objective or in order to establish the directions for a transition toward scenarios in which it could be implemented. To conclude this chapter, we wish to outline a possible framework in which to operate. This framework can be described in terms of a series of tasks and possibilities to take in consideration:

- the possibility to include other layers of data in the observed infoscape;
- the possibility to host the expressions and relations of non-human agents in the Relational Ecosystem;
- the possibility to take in consideration the existence of multiple ecosystems or of multiple versions of the ecosystem.

The first item deals with some of the notions which have already been introduced in Chap. 9: The possibility to give evidence of the fact that the ubiquitous infoscape includes multiple types of data, information, and knowledge. In this book, we have constantly stressed the interest of the microhistories of the city, expression of the city's daily life. But many more types of origins of data and information exist at administrative, bureaucratic, statistic, infrastructure, and transactional levels, which can also be aggregated to represent micro- or macrohistories at all of these levels. These sources contribute to describing the context in which the daily life and evolution of the city take place. Certain regulations make certain actions legal or illegal, possible or impossible, suggested or contrasted. Bureaucracies add to the complexity of performing certain actions or enable the fact that other ones can be performed only in specific ways. Statistics describe the general context in which things happen in the city, giving information about wealth, health, traffic, energy, education, and more. Data about infrastructures can describe why a certain activity is possible or impossible in a certain part of the city, or the reason why it is discussed or absent. Is there free Wi-fi connection in a certain neighborhood? Does this area of the city have water or energy issues? What is the quality of service of road maintenance in this zone? How do these things reflect on what people are talking about? These are only some of the possible intersections.

This type of process can be systematized, for example, by using the Human Ecosystems or one of its functional equivalents. The information environment created by Human Ecosystem can, thus, become augmented by hosting all of these data sources (Open Data available for the territory, administrative, and bureaucratic data, statistics such as census and land use, zoning, land registries, infrastructure data regarding communications, transport, energy, and more). Some of them will be geographically characterized and, thus, will be directly viewable on maps. The others will regard the whole territory as a contextual parameter. All of them will vary over time and, thus, in visualizing the data coming from a certain time frame they will need to be correctly selected.

In all of this process, the risk of the emergence of false correlations will manifest itself and be experienced multiple times. Correlation does not imply causality. Only because two things change proportionally, it does not mean that they are one the cause for the other. This is a frequent statistical phenomenon, and a flaw of many famous statistical researches.

For this reason, these data sources must not only be validated, but also checked for interoperability (they must be homogeneous in their measurements and compatible in the philosophy in which they were captured), and each result must be verified against the incurrence of this phenomenon.

The second item—the possibility to host the expressions and relations of non-human agents—addresses the possibility to interconnect objects and places to the network and to capture data from them. Buildings, plants, devices, appliances, spaces, cars, and other objects can have sensors built in and be connected to the network. In this way, they can become data and information producers and, thus, active and participating members of the infoscape. These objects can be geo-references, so that they can be positioned on maps and other geographical visualizations. They can be given identities (e.g., using the Ubiquitous Commons, as introduced in Chap. 4 and detailed in Chap. 13). They can express their status across time, as the recording from their sensors varies with the situation. They can describe relations as people use them, come near them, interact with them, and more. Thus, they have all the characteristics they need to take part to the infoscape and to the Relational Ecosystem. The same cares dedicated to the data highlighted in the previous point (false correlations, compatibility, homogeneity) must be applied in this case, as well.

The third item—multiple ecosystems—instead—deals with a different type of opportunity. In effect, it would be naive to suggest that a single Relational Ecosystem exists in a city. Even if we were to adopt the most holistic of approaches, to take into account a single ecosystem would be misleading.

To understand why, let us make a practical example: companies. Companies host multiple hierarchies. From the most formal ones (e.g., CEO, management, investors, business units, unit managers, employees, externals, and all of these host yet further hierarchies) to the most informal ones (e.g., the CEO went to school with one of the externals and they know each other well, for this reason the CEO discusses some of the issues in the company with his friend and also values his advice in making complex strategic decisions). A wide gradient of possibilities runs between these two extremes.

If we look at the Relational Ecosystem of the city, how could we represent such a variety?

In the first instance, we could imagine to use administrative and legal definitions. The company is a legal entity and, thus, a recognized identity. For this, we could imagine to add the company as a non-human participant in the Relational

Ecosystem, related to all of the other people and hierarchies through appropriate links.

This would be a good approximation, but it would fail in capturing the different logics with which all of these relations adopt and in modeling and describing the ways in which they function.

Our best option would be to manifest the company and its participants as a series of different identities with a variable degree of humanness (but this can also be of marginal interest), representing the various hierarchies and the members which participate in them. People operate within their hierarchy in different ways compared to the ones which are used toward other part of the company. The company behaves differently than its participants. Formal relations happen in ways which are different from the informal ones. Members of the same company could have relations and interaction modalities which have nothing to share with the ones they have within the company. The company and one of its employees could behave in entirely different ways with the same external subject.

By taking in consideration the fact that there are multiple ecosystems (or different versions of the ecosystems), we become, instead, able to solve all of these problems, by turning each identity in a constellation composed by multiple identities and combination of identities (as suggested by Canevacci's multividual). Even more this constellation would depend from the points of view from which we choose to observe (and to participate) to the ecosystem, and thus, there may exist multiple version of them (and, thus, the multiple versions of the ecosystem).

Just as it applies to companies and organizations, this kind of reasoning could apply to different types of subjects, human and non-human. In general, it can apply to all collective identities in the system, as they establish and relate with other ones according to different logics.

In this chapter, we have taken in serious consideration the possibility to include non-human actors in the Relational Ecosystem, arriving at the definition of a framework with which it is possible to operate such augmentation. In the next chapter, we will see an application of this concept using a specific case study coming from the city of Sao Paulo in Brazil, and then, we will move on to the final chapter before the conclusions, exploring the transformative philosophy implied from Digital Urban Acupuncture: the passage from the idea of consensus to the one of coexistence.

11.2 An Interview with Pier Luigi Capucci

Pier Luigi Capucci *is an Italian theorist, writer and an educator in the fields of media (both in communication and art realms), of the relations between culture, science and technology, as well as an active contributor to the international debate about arts-sciences-technologies. His research and books "Realtà del virtuale. Rappresentazioni tecnologiche, comunicazione, arte" (Reality of the virtual. Technological representations, communication, art), "Il corpo tecnologico.*

L'influenza delle tecnologie sul corpo e sulle sue facoltà" (*The technological body. Technologies' influence on the body and its faculties*), *and* "*Arte e tecnologie. Comunicazione estetica e tecnoscienze*" (*Art and technologies. Aesthetic communication and technosciences*), *position him as one of the most prominent experts in the connections between arts, sciences and technologies, on the shifts and sociological impacts from real to virtual and in the passage to post-humanity. He is a pioneer in digital and networked publishing.*

Estimates say that anything from 8 to 20 % of all social networking activity is generated by software agents, Bots. This does not happen only on social networks, but also on other modalities, on the web. (According to Incapsula's 2015 Bot Traffic Report; 14 % on Twitter, for 23 Million users; 56 % of Internet traffic; 8.7 % of Facebook users) This means that a large part of the information which we experience is not generated by human beings, but from other types of entities, with their logics, attitudes and behaviours: from advertising agents, to weather announcement bots, to spiders harvesting content, to analytics, to malware, to creative, artistic and weird ones, and more. These entities perceive us, too, as demonstrated by their capability to feed us with personalised messages, often with high levels of intelligence (or at least adaptability) involved.

What is changing in our perceptive, cognitive and relational scenario in this sense?

Although the vast majority of people is unaware of it using devices and services, we must accept the fact that a consistent part of our lives deals with—or is ruled by—entities created or activated by humans, but that are not human. Over 50 % of circulating emails is spam, often generated by bots. An algorithm writes the "breaking news" for the *Los Angeles Times*. An Artificial Intelligence program predicting the success of investments is "sitting" in the Board of Directors of a Hong Kong company, with the same right to vote as its human fellows. The fact that it is considered so important to be included as a peer in the governing council of a company, means that its predictions have proven to be generally more reliable than the ones of its human colleagues. Increasingly efficient voice assistants are now part of smartphones (Apple Siri, Microsoft Cortana, Google Now are the most popular), paving the way for vocal interfaces. And projects and prototypes of self-driving cars proliferates, and according to analysts will spread within a few years. Famous players such as Mercedes, Google, Toyota, Nissan, Tesla and, apparently, even Apple are already committed to it. This scenario, and the fact that these vehicles of all types, sizes and modes of travel will use non-polluting energies, will change the near future of mobility. Fundamental issues of software security must be addressed and solved, not only about privacy but also, obviously, about violability—due to design errors, bugs or more or less legal and unauthorized intrusions (by crackers, hackers, judges, police, etc.). A group of security researchers has violated the system of a prototype vehicle thanks to its Internet connection, turned off the brakes and manipulated the internal vehicle services, so it is urgent to find concrete solutions. If these problems are solved, self-driven cars will be less harmful and safer than the human-driven ones: they would fully respect the traffic laws, and insurances would be cheaper. Owning a vehicle would make little sense since, if necessary, we could subscribe to transportation services that more widely than current taxis could reach us and take us where we want to go.

Typical in human history is the tendency to delegate tasks, labors, repetitive and unattainable operations to external devices and entities, created by humans and informed by human knowledge both in their making and in their working—and perhaps even in their long term perspectives. Typical is also the attempt to build living-like entities, simulacra of human and animal life able to raise humans to deity, to the Myth of creation. Examples are countless: the Greek Talos, the Hebrew Golem, the medieval, Renaissance and eighteenth century automata, the nineteenth century Frankenstein, the twentieth century robots, androids, cyborgs and replicants largely described in literature, the sophisticated social and industrial contemporary robots. "Robots" are historically present in almost all cultures, from Europe to Islam, to China, to Japan. A book on medieval robots was published earlier this year[3], and from October 2015 until February 2016 ZKM is organizing in Karlsruhe a major exhibition on medieval Arabic automata, with a rich conference program[4].

Quoting Hans Moravec, the prominent scientist in robotics, these more and more efficient and autonomous entities, devices and machines generated by human culture, are "the children of our minds", which is also the title of his famous essay (*Mind Children: The Future of Robot and Human Intelligence*[5]). In his book Moravec refers to robots, but softwares are human creations too, they are "children of our minds" as well, and today a significant part of our lives and habits is based on softwares. These increasingly complex, autonomous and flexible entities can replace us in many specialized and repetitive tasks, finalizing them better than what we could do, and they can participate to missions that are simply unattainable for us (on Mars there are robots, not humans). Their skills are often vertical, namely they know how to do a few things—sometimes just one—but they repeatedly and quickly work without getting tired and with no errors: in fact they could be considered as new categories of slaves. By now, these dark and obedient entities—pervasively, openly and discreetly acting in almost all devices and environments of everyday life—are doing in a disciplined manner what they have been ordered and trained to do (I doubt that the above mentioned Artificial Intelligence has been asked to enter that Board of Directors). So far they have supported and enhanced human activities, but soon they could take autonomous decisions and independently develop many skills, they could express volition, self-organization and replication capabilities: they could *evolve* behaviours similar to living organisms. This scenario, made more complex and general by a parallel evolution in the fields of Genetics, Synthetic Biology (the organic life forms created or modified by human culture), Robotics and organic/inorganic hybrids, would mean the emergence of

[3]Truitt (2015).

[4]Globale: Allah's Automata. Artifacts of the Arab Islamic Renaissance (800–1200), 31/10/2015–28/02/2016.

[5]Moravec (1988).

what I have called “Third Life”, being “First Life” the biological life and “Second Life” the life in the symbolic dimension[6].

Software agents and complex non-human entities influence our behaviour, politics, economies and perceptions. We can appreciate just how much simply by looking at the stock markets (which are mostly driven by automatic software agents), or by the social experiments which subjects like Facebook continuously conduct on millions of users at the time.

In your opinion, and starting from your research, what are the most characteristic or recurring features of these types of phenomena?

A software is ultimately a set of instructions with a goal: a powerful language of symbolic derivation, since it has been conceived by humans, created to operate a variety of devices, to govern in a simple way the complexity of a device’s activity and of the interaction with it. Although we often realize it only *ex-post*, the tools and devices we create and adopt deeply influence our behaviour. Sticking to the contemporary era, the devices raised by computer science profoundly affected Western societies, the ways of living, working, learning, playing, getting in touch. Like in any major change, new jobs and professional figures emerge, while others are put aside.

Today software is ubiquitous and relevant in our everyday life, it is hard to escape its pervasiveness. In my opinion there are two kinds of problems, of “passive” and “active” type.

The first case is the possibility to track behaviours, ideas, information about us, for purposes we are partly aware of (informing, advertising, selling and buying, communicating, sharing, having fun, meeting, etc.) or completely unaware of (analysing behaviours, profiling habits, retrieving economic, familiar, medical, information, etc.). Today with the vast amount of data about us that is available from many sources it is possible to create comprehensive profiles that can be used for various purposes we don’t know. The issue of privacy protection is absolutely central in our society, despite it is slightly perceived by the public opinion. In fact, on one hand the sacrifice of privacy is coupled with an increased personal and social security: people are willing to accept an increasing control on their data by some authorities in exchange for a greater security, since they think of “doing nothing wrong” (the US Congress has recently approved a law that allows authorities to access smartphones, tablets and computers contents, officially to combat crime[7]). On the other hand, we often enjoy spreading private information, activities, preferences about ourselves, for example on social networks, ignoring that somebody could see and use them for purposes that are totally unknown to us, and even unacceptable or fraudulent.

---

[6]Ascott et al. (2009).

[7]Greenberg A, Grauer Y (2015) Cisa Security Bill Passes Senate with Privacy Flaws Unfixed. Wired, 27/10/2015, http://www.wired.com/2015/10/cisa-cybersecurity-information-sharing-act-passes-senate-vote-with-privacy-flaws/ Accessed 28/10/2015.

The “IoT—Internet of Things” represents a further step. In its most extended execution it provides the assignment of an IP label to every artefact: objects become recognizable and acquire the ability to exchange data with other objects, thus taking an active role thanks to the Internet and creating a pervasive and interconnected system. In an IoT environment, the virtual world draws a map of the physical world giving an identity to things and places. However the IoT could ultimately become the most pervasive and effective control system of the territory ever *evolved*—in this case *invented*—by a species, since the next step beyond its use in urbanized areas will be the interconnection of the natural environment. There are already applications to monitor earthquakes, weather, fires, landslides, avalanches, rainfalls, vulcans’ activity, climate change, forests, animals migration, species extinction… The control of the environment would be more pervasive, sophisticated and refined thanks to digital, optical, bio-and nano-technologies. For all living species, hence for humans too, the control of territory is crucial. Human species uses artificial technologies for this task, creating a contradiction between the legitimacy of biological dimension (and of environmental monitoring) and privacy protection.

Besides privacy issues, the second kind of problems has to do with the “active”, autonomous and dynamic capabilities of software to assemble large amounts of data from many sources, generating overall scenarios that can show macro-trends (commercial, economic, political, social, cultural, etc.). These complex scenarios—generated by software using variables that can be introduced or omitted, on purpose or by ignorance—can deeply influence public opinion and politic choices.

In 1995, at a conference on the right to communication in the era of software and media technologies, I argued that the boundaries between democracy and social control would have been increasingly feeble, that the risk was a “democracy of enchantment”—which was the title of my speech—both in the meaning of a democracy of “fascination” and in the meaning of a democracy of “immediacy”, of “impulsivity”.[8] I think this is precisely what is happening today. Therefore, we must act to prevent this drift, but how? By establishing an “over-controlling” entity, from the above? Through a diffuse and widespread control with “switches” that can be turned on or off by people, from the bottom? By releasing free, accessible software which people can manipulate, eliminating secrets and closures? Or do we need to rethink the very concept of “democracy”, its aims, limitations, rituals, hierarchies, functions, as well as the conditions for its implementation? A further demonstration of how technologies can influence and shape the reality where we live and where they operate.

Even if it is human beings who design and develop the logics and infrastructures for these agents, whether we understand the implications of their “life” and actions is not certain, nor a simple thing to understand. Complex algorithms, neural networks, artificial intelligences are creative in more than one way. For example, regarding our behaviours, software agents continuously classify us through variables which they create themselves, which we don’t

[8] Democrazia dell’incanto (1996)

understand and for which don't have names. We might think, for example, of the variables used inside a neural network designed to classify social network users. Or phenomena such as the Flash Crash of the stock market in 2010, where major newspaper mentioned "mysterious algorithms".

What are the implications of dealing with complexities which we don't and can't understand?

As I said, the logic of action of these "agents" is largely hidden or incomprehensible to the majority of people, and often even to those who manage them. However, in principle, understanding how the tools and devices we use operate is not mandatory. Many people do not know how the combustion engine of a car or the electric engine of a train work, they have no idea what a microprocessor or a video card is, but that does not stop them from driving a car, travel by train and use a computer, that is living reasonably well without knowing all this stuff. We must not, nor evidently can, and sometimes not even want since we are not interested, be experts in everything we use. In Computer Science the development of graphical user interfaces and the rise and spread of personal computers ("the computer for the rest of us", that is, for non-experts) are based on this. Even people who have to take decisions with a relevant social and political impact, such as signing laws and thus affecting the citizens' lives, are not commonly expert in the fields they are intervening on, although it could be helpful. Indeed, different levels of expertise are needed to solve these problems, and the extraordinary symbolic human ability, which enables people to collaborate by sharing ideas, skills and knowledge to achieve results that go beyond the mere sum of parts.

This situation is different when dealing with entities capable to autonomously modify themselves, which tend to escape to control and understanding (sometimes even of the experts), which behave and act in complex ways that are similar to the behaviour of the living. These entities could eventually escape the tasks and purposes they were made for; or they could be covertly modified to generate new information and knowledge for subjects who are not the ones who have made or commissioned them; or these entities could, more or less deliberately, introduce errors. By now our culture has only a few defences to face that increasingly independent and complex New Nature it has created: in the information technology field, unplugging devices and cutting off energy is a radical, rough, obscurantist and unlikely solution. Anyway, we have to find ways to introduce behaviour rules—a sort of ethics—in these complex devices and applications generated by human creativity. In Robotics the three famous Asimov's laws can be considered a simple attempt to give to machines a kind of basic human ethics. Laws that—with increasingly sophisticated robots that will spread soon in the social realm, with behavioural responsibilities not ascribable to their owners—are to be revised and expanded. That's what is doing Roboethics[9], working also in the industrial and military fields.

---

[9]Tzafestas (2015).

Even laws, which usually moves at a very slow pace, are starting to recognises these entities. Some are already established in our perception and legal systems: companies and organisations are legal entities, with personality, identity, addresses, possessions, property, rights, intentionality, responsibility. Some of them at a slower pace, and in ways which are more dubious.

What are the evolutions in this sense? How are legal systems and people reacting to issues like these, which are not only technical, but also philosophical, social and political?

Law sector has been deeply shaken by the advent of information technology. Legal systems move very slowly, often in a confused and retrograde manner: controversial laws on software ownership, on piracy and copyright, on the distinction between copy and original, are exemplary. This distinction does not lie inside software but in external certification elements (SIAE stamp in Italy, presence of holograms, notes on the packaging, creation date, etc.). In fact, in the digital realm there is no difference between copy and original. What's more, if there is no data compression or reduction, the distinction between "copy" and "original" does not make sense, since the "copy" is structurally and functionally identical to the original. Software is not something material, you don't own it, you only have a license to use it, which sometimes is limited in time and can be revoked. You can disassemble and analyse the engine of a car, but disassembling and analysing proprietary and commercial software is an infringement of the license agreement. Despite its indefinite duplicability, it is prohibited to copy proprietary software: its user license is personal. Software emphasizes how economic value can not reside only in material objects, but also in the immateriality of algorithms and programs, in the intangibility of processes and procedures. This paradigm shift, totally unbalanced in favour of software industry rather than users (for instance manufacturers are not responsible for possible malfunctions caused by their products), helped to make software and software-based services companies among the richest ones worldwide.

On top of these legal issues, more than thirty years old and essentially based on ownership and software use, there is the recognition of, shall we say, "social responsibility" of software, until now dependent on the responsibility of users and sometimes of manufacturers, who eventually bear legal accountability. A greater software autonomy would require a legislation which takes into account its responsibilities. If the Artificial Intelligence I mentioned before, "sitting" in the Board of Directors of that Hong Kong company, fails its forecasts with the result of wasting money and bankrupting the company, would its creators or its managers bear the only responsibility? Or, rather, would the responsibility be in the creative skills though which that A.I. invents new ways of reading and interpreting the data?

New legal issues in other technological contexts related to softwares begin to peep. Suppose you leave your car on a sloping road forgetting to engage the gears and without pulling the handbrake, and suppose that the car ends in a cross causing an accident. Who is responsible for the accident, the car or the driver? The answer is obvious: the driver and his/her forgetfulness. The car is considered an inert entity, running blindly its driver's commands. But if, instead, the accident was caused by an advanced robot, autonomous and self-sufficient, capable of volition or even consciousness, able to decide its actions, the question would be very different. It should be necessary to develop an *ad hoc* legislation, which recognizes the responsibility of the robot—and therefore also its legal status, duties and rights. The robot would be a new type of legal entity, different from humans but also from other "inert" machines, a sort of halfway entity. Roboethics is trying to address these topics as well.

One thing is certain: that we find ourselves already within an ecosystems together with non-human entities: from microbes, ants and cows, to corporations and artificial intelligences. Maybe these last ones are not similar (yet?) to the ones shown in Hollywood movies, but they are already a consistent part of our reality.

How do we communicate within this type of ecosystem? How can we exchange data, information, knowledge and wisdom with these entities, in ways that are meaningful and, potentially, mutual?

True, little by little we realize that our reality is far more extensive and complex than we thought not so long ago. It is a wide panorama populated by very different subjects. In addition to the (more or less known) ecosystem made of plants and animals, there are inorganic immaterial entities, such as Artificial Intelligence, Artificial Life, software and algorithms; there are inorganic material entities, such as increasingly sophisticated machines, devices and robots; there are natural, organic-based organisms such as bacteria, of which we have recognized the importance and influence on our lives; there are artificial organisms obtained by modifying natural subjects; there are artificial organisms created from natural subjects, like in the case of *de-extiction*, whose aim is to revive extinct species by selecting and cross-breeding existing species or by genetic engineering; there are artificial organisms newly created through Synthetic Biology and Genetics; there are hybrid entities, which combine organic and inorganic elements, as it is happening in the field of bio-robotics. This list is intended to lengthen.

Some of these organisms/entities have always existed, but only recently we started understanding their relevance: the case of bacteria is exemplary. Other organisms/entities are the recent result of human symbolic and design capabilities, and in simpler forms they have always populated the history of humanity. I have called "Third Life" the new organisms/entities created by humanity.

How can we exchange information or cohabit with these subjects? Some of them, like bacteria, are problematic only in case of pathologies, since natural processes are generally self-sufficient and autonomous. With vegetal and animal ecosystems we constantly interact although not always in positive ways, acquiring increasingly responsibilities on the state and maintenance of the natural ecosystem. With artificial organisms/entities things are more problematic, because they are human artefacts created for purposes which can be very different, distant or competing, sometimes invisible, elusive or incomprehensible to most people, unless adopting suitable technical and cultural tools. Things are more problematic also because the evolution of artificial organisms/entities is far more rapid than human biologic and cultural evolution, making harder to adapt to such fast processes. This acceleration is increasing.

Sometimes non-human lifeforms have proven to poorly adapt to human communication contexts. A frequent experience is interacting by phone with synthetic voices used in automatic answering machines or voice systems, now common in contact with large companies. I think that these interactions have often an incomprehensible and confused logic, and I feel irritation when non-human entities acting as humans refer to me, sometimes not so gently. If you want some of my time and attention to answer questions that are mainly relevant to your business, you must have the respect to do it with a real person, or inventing an intriguing communication. As Michael Dertouzos, former professor at M.I.T., states in his book *The Unfinished Revolution*[10], a few tens of dollars device undermines an invaluable apparatus like the human brain. To avoid misunderstandings and prejudices is therefore necessary to improve these new (and not so new) communication forms between human and non-human subjects.

As to the communication among humans in this ecosystem, many devices are born with the function to communicate, share and exchange information, although not always in a correct and social way. Symbolic communication is the "genius" of our species. The history of humankind, from the acquisition of symbolic ability, could be considered as a formidable set of attempts and constructs to communicate increasingly faster and further, in more and more effective, simpler and cheaper ways. An obvious example is Skype (and similar services/platforms): before Skype only TV broadcasters were able to implement remote audiovisual communication in real time (the "live broadcast"), it was a very costly task managed by few. Today, a significant portion of the world's population is able to perform an audio-video communication process in real time almost for free (using smartphones, tablets or PCs and accessing the Web).

This scenario opens up new types of conflict: between human beings; between non-humans (see, for example, the current "algorithmic" dispute between Youtube and Chrome, both owned by Google, in which Chrome's strategy aims at defending browsers from advertisements, while Youtube's aims at placing ads in subtle, unexpected, user-uncontrollable ways, potentially influencing billion dollar markets); and among the two.

For example, it is interesting to observe the strategies according to which certain corporations fund specific research areas. What happens if research is funded to debunk climate change? Or in favour of tobacco smoke (as it happened in the previous century)? A human or non-human decision?

What are the new forms of conflict which we will see rising, in this sense, in the next years?

Conflicts have always existed in human societies and they took advantage of all the available tools. Conflicts are not always necessarily negative: they are often at the basis of human societies, democracies and governments.

Today some of the conflicting issues concern the new organic, inorganic and hybrid entities created by human culture, their destination, their possible employments (commercial, industrial, social, military…). And also their evolution in human societies, their role as active subjects—as living in the living—is to be

[10]Dertouzos (2001).

determined, both socially and legally. As I argued before, humans create new artificial organisms/entities to essentially create slaves, able to carry out tasks that are repetitive, that we don't want to do or we are not able (or not so able) to accomplish. Since the past societies were slavery was in force were violent and warmongering because of the low value placed on human life, some scholars assert the possible emergence of societies dominated by violence and war. Today, drones are used in military operations, robots and synthetic or genetically modified organisms might be used soon too. Hence an ethic of the "Third Life" seems to be undelayable.

In our daily life, an invisible infosphere exists alongside the visible one, an information flow that is inaccessible to us although it concerns and influences us, which we are only partially aware of and unable to fully control it. The risk is to become the objects of this dark infosphere, without knowing the principals, the origin, the reasons, the destination, the boundaries, the goals. Radicalising what Cyberpunk addressed in the 80s of 900, today there are no more places that may escape communication, not even the deserts, the Poles, the most inhospitable areas on Earth: it is as if symbolic communication, the "genius" of humanity, had turned against us. We can no longer be "outside" because any "outside" is always in some "inside". We are forced to some degree of inclusion, and all what we can do to avoid adaptation is trying to use the same tools to react, creating temporary spaces, impermanent territories, transient cycles: being "in between", in the passage, where analysis, profiling, cataloging we are constantly undergoing get confused, uncertain, unreliable. The New Nature, beyond embodying and unfolding the power of the possible and the wonderful, can also produce a sensation of fake, deceitful, deceptive, uncertain, and even disrespectful, aggressive, irritating, violent. Maybe the millenary conflicts, the fanaticisms and fundamentalisms that are strongly (re) emerging today, are related to the search for a sort of an illusory restoration of values, beliefs, rituals, roots, to preserve and counterpose in order to build new foundations, to desperately create some identity in which being mirrored. Although the underlying reasons have dried up long ago.

References

Ascott R, Bast G, Fiel W. Jahrmann M, Schnell R (2009) From life to life. The multiplicity of the living. New realities: being syncretic. Springer, Wien

Dell'incanto Democrazia (1996) Democracy of enchantment: in Strano Network (ed.), Nubi all'orizzonte. Diritto alla comunicazione nello scenario di fine millennio (Clouds on the horizon. Right to communication in the scenario of the Millennium's end), Rome, Castelvecchi

Dertouzos M (2001) The unfinished revolution: human-centered computers and what they can do for us. HarperCollins, New York

Moravec H (1988) Mind children: the future of robot and human intelligence. Harvard University Press, Cambridge

Truitt ER (2015) Medieval robots: mechanism, magic, nature, and art. University of Pennsylvania Press, Philadelphia

Tzafestas S (2015) Roboethics: a navigating overview. Springer, Wien

Chapter 12
A Case Study: São Paulo and Its Real-Time Museum of the City

Abstract A Real-Time Museum of the city of São Paulo, dedicated to creating an iconic, engaging, inclusive experience of the Relational Ecosystem of the city, built together with SESC, the Brazilian federal agency in charge of managing the nation's museums and cultural installments. The museum (which is currently being built at the time of writing) has been designed to create awareness about the many layers of the Relational Ecosystem of the city, and to provide a cultural shock to the population, to promote a radical transformation of the sense of possibility about the perception of what a metropolis can be and about the ways in which the concept of citizenship can be transformed, with citizens becoming positive, active, aware, and participatory agents of the city's civic life. The museum hosts a laboratory in which all members of society can learn how to use the Relational Ecosystem of the city, using Digital Urban Acupuncture techniques, to pursue their goals and to create patterns of self-organization and intervention on the urban environment, supported by the city, regional, and federal administrations. The chapter narrates the methodologies, design process, objectives, scopes, and expected impacts of the project, as well as documenting the patterns, expectations, and future scenarios for this initiative.

12.1 Introduction

Sao Paulo is an immense city. Its metropolitan area is a giant agglomeration of 39 municipalities which contain around 17 million inhabitants. The city produces a relevant part of Brazil's entire gross national product. Since the introduction of the car, Sao Paulo's expansion has followed a radial model: from 1 km in 1870 to more than 80, today. Land use is characterized by high levels of speculation, leaving behind wide areas of underdevelopment and difficult scenarios, and accentuating social segregation.

The dataset for the Sao Paulo case study is available at http://www.human-ecosystems.com/DAP/DigitalUrbanAcupuncture-CaseStudy-SaoPaulo-dataset.zip.

S. Iaconesi and O. Persico, *Digital Urban Acupuncture*
DOI 10.1007/978-3-319-43403-2_12

The city is also a place of continuous, radical cultural transformation, and mutation. Brazil's cultures constitute an exceptional case of constant remix, perpetually recombining traditional, avant-garde sources from within the country and from throughout the world, describing an ever changing scenario of mutation. This is visible at all levels and dislocations of societies and communities: Brazil and, in particular, cities like Sao Paulo are attractors of practices and approaches from all over the world, which here find spontaneous and emergent integration, bringing change, and evolution to society and cultures. This does not happen without conflict, as the transformations in music, visual arts, and practices of consumption (anywhere from nouvelle cuisine to drugs) highlight single or general confrontations which are present within and throughout the communities.

Sao Paulo is also a city of profound paradoxes, where the most interesting innovations can happen in favelas as well as on the rich and financially luxurious Paulista Avenue, with the same relevance and cultural impact.

From jazz, bossa, to electronic music, to hardcore metal, to traditional music from the Caribbean, Africa and the southern american continent; from capoeira, to breakdance, to ballet, to contemporary dance; from brutalist architecture, to modernity, to generative architecture, up to the intricacies and informal labyrinths of favelas. All of these are just some of the examples of the complex cultural reality of Sao Paulo, each element contaminating the other.

Sao Paulo is the communicational metropolis, to use Massimo Canevacci definition, a city in which fluid, liquid, recombinant communication pervades all spaces, becoming the city itself, more than buildings and streets. In this setting, continuing the series of paradoxes which characterize this place, this is also the city of "Lei Cidade Limpa," the Clean City Law, the urban regulation from 2007 which has banned all forms of outdoor advertising, including signs, billboards and urban screens, to fight visual pollution in the city. And it is also, at the same time, the city of urban projections and media facades—for example with the FIESP building with its 3700 m^2 led display—in which arts and design merge to provide new communicational and interactive experiences of the city and of its life.

Sao Paulo is also the city of conflicts, in multiple directions and senses, from the financial elite, to the extensive favelas, to the rising gentrification, to the walled gardens of the many gated communities, to the rise of unemployment, to the struggles between governments and citizens, between corruption, violence and political dissonance.

A different form of conflict, ranging through a series of degrees of arousals, from quiet and peaceful coexistence of different cultures, up to violent uprisings and riots, is determined by the compresence of a wide variety of different cultures in the city, particularly evident for the presence of the many "embedded cities"—such as Sao Paulo's Chinatown, KoreaTown, Japanese neighborhoods, Quilombos and Mocambos, the difficult situations of the natives, and the incredible mix of cultural remixes which come in any neighborhood, opening the city up to colors, attitudes, behaviors, desires, expectations and visions of the world of any kind.

Here, in Sao Paulo, we were invited by the Universidade Metodista and from SESC—the Serviço Social do Comércio, a nonprofit, private institution which is open to the general community to provide innovative welfare solution under the form of wide actions and by creating large, disseminated centers for art, culture, participation, health and well-being—to create an instance of the Human Ecosystems.

In particular, the Universidade Metodista, with the interest of Fabio Botelho Josgrilberg, at the Social Communications Program, was keen to understand how the Human Ecosystem's concept and the Digital Urban Acupuncture methodology could be applied in a city such as Sao Paulo. At the same time, SESC wanted to explore the opportunities provided by HE and DUA, to create a novel form of museum, which could also become the epicenter of wide, inclusive processes for civic participation and action, and to diminish the digital and cultural divides which are present in Brazilian cities.

12.2 In Sao Paulo

When we arrived the first time in Sao Paulo to set up what would then become the Human Ecosystem in 2012, we were immediately overwhelmed.

We had already been in the city before. But this time it was different. We were about to start a research project whose objective was to gain better understandings about the public spaces of the city, by using social media. Thus we were on the lookout, alert, in a state of constant aware observation.

As soon as we left the airport the city presented itself to us through a massive traffic jam in which the smells of pollution and the expressions of agitation and anxiety were more than palpable: from the movement of the cars across the lanes of the highway connecting the Guarulhos airport to the city; to the facial expressions of drivers in the other vehicles.

When we reached the city, heading for our hotel, the city continued to assault us through its features. Extreme differences in wealth, pollution, the jungle of concrete, the vegetation that seemed to be pushing in real time to grow out of the constraints of bushes and flower beds. People also performed a polyphonic and dissonant symphony, ranging from busy businessmen, homeless people, citizens minding their own daily routines, up to the dwellers of bars, relaxed and calm, chatting with their table neighbors, or simply watching people go by.

By the end of the journey from airport to hotel, we had changed our emotional state dozens of times, turning from curiosity, to fear, happiness, anxiety, sadness, surprise and more.

Some of these changes were due to the urban landscape. The alternating of skyscrapers, favelas, modern architectures, postmodern in-betweens, bars, worn-out buildings, parks, squares, and their conditions suggested emotional voyages with their sequence and characteristics.

On the other hand, our emotional changes were prompted by city dwellers, their presence or absence, their behaviors, their activities, and expressions: Beautiful squares could become scary; awful street corners could turn into happy spaces for colorful bar conversations; highway underpasses could become places of joy or fear depending on the context, population, and time of day; parks may seem shady or suspicious depending on who was there, what they seemed to be doing, or through the absence of people. Streets, skyscrapers, sidewalks, and even sculptures, signs, and shopping centers completely changed for us, according to how city dwellers occupied them, what they expressed.

After a few days, our perception of spaces in the city radically transformed.

Progressively, we became able to read additional layers in the city's landscape, recognizing patterns, gestures, activities, and interactions which were not readable in our first immersion in the urban environment. We started discovering places which, at certain times of the day or of the week, completely changed their characteristics, due to the presence of different mixes of people, dress-codes, styles of spatial appropriation, facial expressions, and emotional character.

The emotional expression in places transformed the ways in which we experienced them, thus provoking different emotional responses on our side.

It was an interesting feedback loop.

It was a emotio-spatial hermeneutic: "Emotions are understandable—'sensible'—only in the context of particular places. Likewise, place must be felt to make sense." (Davidson and Milligan 2004).

In some partially elusive ways—through representation and other modalities—emotions, places, times, and contexts are interconnected.

We decided to start our DUA process focusing on emotions, and on the ways in which they flow throughout the Relational Ecosystem of the city, to describe its dynamics and stories.

We set forth to ask ourselves whether it would be possible to use emotional expressions on social networking services to gain better understandings about the emotional expressions in urban spaces. And, if it was, what kind of insights would we be able to gather through this practice, and how could they be used.

12.3 The Sensing Framework

We have, thus, set forth to capturing data from social networks being generated in the city of Sao Paulo, to use it to grasp geographically locatable emotional expressions, looking for the possibility to identify patterns and recurrences across time, space and context, and how they would interweave and spread through the human network of the Relational Ecosystem.

Many models and approaches are available to evaluate emotional expressivity.

Desmet (2002) has described a basic model of emotions—on top of the work of Rosemann (2001), Ortony (1988), and Lazarus (1991)—in which emotion is the result of an appraisal process fueled by stimuli and concerns.

In this model, the concept of appraisal is defined as a nonintellectual, automatic evaluation of the significance of a stimulus for one's personal well-being. The act of emotional appraisal leverages the presence of two different inputs: stimulus and concerns. According to Frijda (1986) a stimulus is any perceived change which has the potential to elicit an emotion. Every emotion hides a concern, that is, a more or less stable preference for certain states of the world. Thus, (Frijda 1986) concerns can be regarded as points of reference in the appraisal process. It is interesting to note concerns' relation to Maslow's (1943) hierarchy of needs: Concerns can be placed at the different levels of the hierarchy to express their referring domains, their importance, and priority.

Some models directly deal with the possibility to interpret emotional expressions and responses in architectural contexts.

Ulrich (1983) and Berlyne (1971) have focused on complexity as a parameter for evaluation, referring to the number, and the varieties of different units present in a setting/picture.

Kaplan (1987) proposed a model on environmental preference in which predictors like complexity play an important role. According to Berlyne (1971), different levels of complexity can be associated to different levels of preference.

According to Zajonc (1980), familiarity and prototypicality influence the emotional reaction to architectural spaces.

Russel (1980), then, proposed the circumplex esthetic model in which the affective appraisal of the environment can be summarized by two dimensions: pleasantness and arousal. This suggested Purcell's (1986) Discrepancy model in which the level of likeableness attributed to an external stimulus depends on how far is the appraised environment from the prototypical exemplar that an individual has in mind. If the incoming stimulus is too similar or too different from the prototype, it is very possible that the evaluation will be negative; on the contrary at a moderate level of discrepancy, there are good chances that the stimulus will be appreciated.

Some other methods focus on the daily life of citizens.

In Nold's Biomapping (2004) and Emotional Cartography (2007) projects multiple people in 25 cities were asked to wear galvanic skin response devices to record their emotional arousal in conjunction with their geographical location. In this way, maps are created which visualize points of high and low arousal.

Wilhelmer's, Von Bismarck's, and Maus' Fuehlometer ('feel-o-meter') project (2008), came under the form of a public face, an interactive art installation that reflects the mood of the city via a large smiley face sculpture. It was installed atop a lighthouse in Lindau, Germany. A digital camera along the lake captured the faces of passersby, which were then analyzed by a computer program and classified as either happy, sad, or indifferent. The cumulative results determine the expression of the sculpture, whose mouth and eyes shift accordingly via a system of automated motors.

Using a different approach, the City of Vilnius (2013) used a social tool to gauge the average residents' level of happiness, using an online form and barcodes in the city.

Another example comes from our Consciousness of Streams project (Iaconesi and Persico 2011), in which art installations in multiple cities were used to contribute users' emotional state and geographical location, creating a "real-time interconnected emotional map of the planet" (Iaconesi and Persico 2012).

The Mappiness project (2011) created a mobile app which contacted users once a day to ask how they were feeling. The resulting data would be used to create information visualizations.

The Aleph of Emotions, an experimental art project by Vigneshwara (2013), used a camera-like interface allowing users to point along a particular direction, focus to a place along that direction, and click to view a representation of emotions in that place. Emotions were collected by searching for keywords on social networks.

Taking inspiration from all of these methods and examples, we set forth in the capture of geo-located user-generated content from major social networks (Facebook, Twitter, Instagram, Foursquare, and Yelp) in the city using a technique based on natural language analysis processes (Iaconesi and Persico 2014), processing them using the LSA (Latent Semantic Analysis) to understand the emotions expressed by the messages, in different parts, times, and contexts in the cities.

The result was a massive database of geo-located, time-stamped user-generated emotional expressions in 29 languages: Between April and October 2014, more than 9 million messages have been captured and classified emotionally.

We used the information produced in this way to try to identify recurring patterns, across time and space, which could provide evidence of peculiar, recurring and/or systematic emotional expressions, and grabbing inspiration from the reference models which we chose to adopt (as explained in the previous paragraphs).

For example, we selected from the messages descriptions of possible stimuli, coming under the form of participation to events, visits to a new location or context, going to the office, and other situations which belonged to users' daily routine or to exceptional situations such as holidays, anniversaries, and situations of high intensity stimulation, such as danger or fear. Then, we paired them to possible expression of concerns, moving back and forth along the content's timeline for each user.

This was possible by using the information produced in the processing stage, using natural language processing to provide us with the topics and emotions expressed in the messages and their sequences and geographical locations.

At this stage we classified these patterns according to the circumplex model, along two axes describing the state of pleasantness and arousal described by each element belonging to the identified patterns, as well as the ones resulting from their aggregations (implemented through k-means clustering techniques, to group information into clusters along the axes).

As a result, we obtained a series of {time, location, topic, and emotion} tuples, grouped within patterns, which we could use for some initial visualization.

In Fig. 12.1, a specific time frame has been highlighted on the interactive timeline to show how the expressions generated in the city at that time distributed themselves in the pleasantness/arousal plane, effectively describing the emotional landscape of the city, as it expressed on social networks. In the image, larger dots correspond to clusters: The larger the dot, the higher the number of people expressing the same emotional positioning. As it can be noted in the image, the pleasantness axis is both positive and negative, with the negative corresponding to

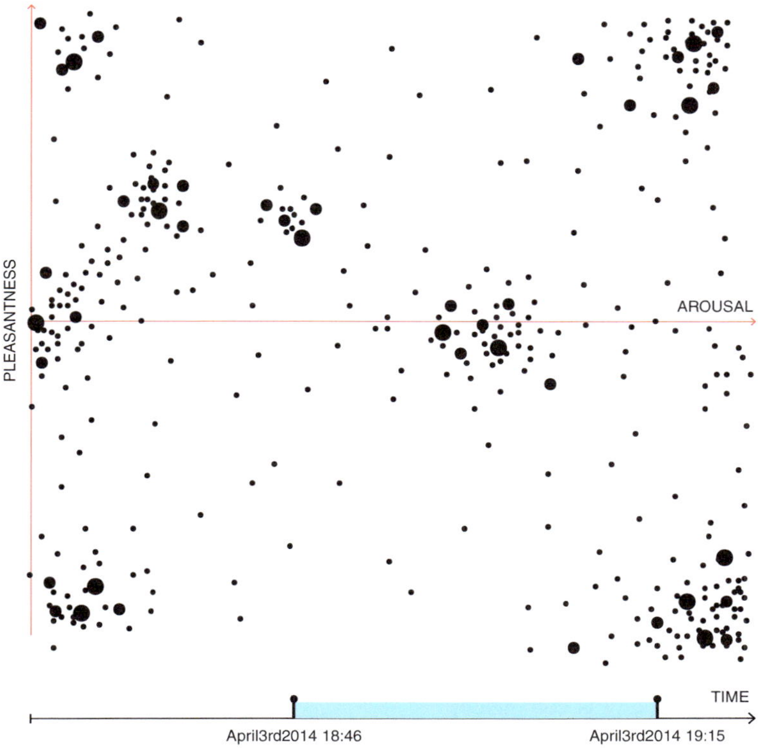

Fig. 12.1 An Example screenshot of an emotional classification of the expressions in the city at a specified time–date interval

unpleasantness, or discomfort. The arousal axis, instead, has only the positive side, with the zero level corresponding to complete calm, and the lowest energy level.

Another example, in Fig. 12.2, shows a geographical distribution, where the green spots of the heat map highlight the locations in which expressions of anger have been published in the city, while speaking of the topic of "jobs."

This clearly shows how coherent areas can be recognized and analyzed.

12.4 Commoning

The data was (and still is) made available in 2 different ways:

- through a series of open data sets obtained from periodic exports (once a month) of the databases which contain the arrays of {time, location, topic, and emotion} for the city, the Relational Ecosystem (a node-link graph, represented using the data structures described in the previous chapters) and a set of aggregated data

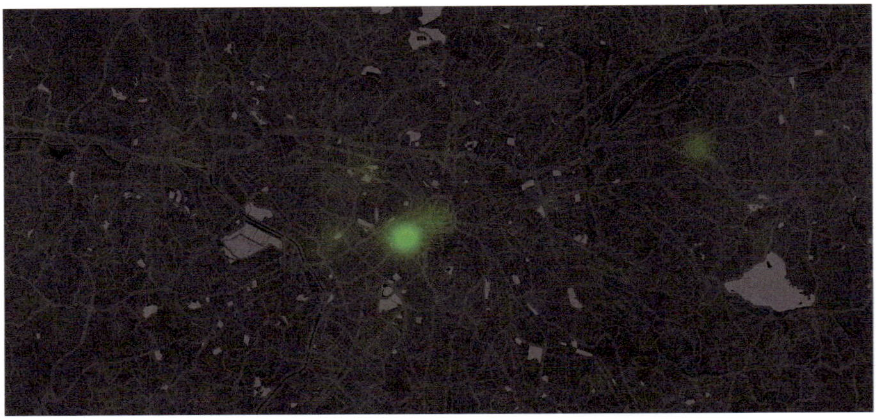

Fig. 12.2 A geographic distribution of the "anger" emotion for expressions discussing the topic "jobs"

which allow to quantify the emotions, topics, discussions, relations, and presence at the various locations at the time of the exports (and, thus, allowing to create time series by juxtaposing the content of different exports); at the time of writing the data sources are hosted on servers owned by Human Ecosystems, until the formal procedure (in progress) for them to be hosted in the Open Data portal of the city is completed;

- through a set of Application Programming Interfaces (APIs) included in the Human Ecosystems software framework which can be invoked manually (e.g., through a Web browser) or by using software products over the network, and which allow to query the databases in multiple ways.[1]

The access to both the datasets and the APIs is regulated through a simple Creative Commons license of the "Attribution, Share Alike, v4" type, which does not impose any restriction on the usage of the information acquired through the system, adding the single obligation to provide Attribution—by specifying in any products, services, and visualizations which may be produced that they have been created thanks to the Human Ecosystems Sao Paulo initiative—and to "Share Alike," indicating that any product which is the outcome of this process must be shared in similar ways.

The "Share Alike" constraint allows to include full accessibility to the knowledge generated through the Human Ecosystems project: when the products of creativity, the applications, the services, and any other elements produced by using Human Ecosystems—and the data captured through it—are shared in this way, it becomes possible to include them back into the resources which are available

[1]The most updated list of these API endpoints can be found at https://github.com/xdxdVSxdxd/HumanEcosystems/blob/master/HE_v2/API.txt.

within the Human Ecosystems, for both the city of Sao Paulo and for all of the other cities which use it.

This means, in practice, that if an artist creates an artwork based on, say, the emotions of the city, or a designer uses the expressions about traffic to create an application for personal mobility—or any other possible usage—all of these must be "shared alike." There is no limitation to what the authors of these works can do with them (including selling them and generating revenues), apart from the fact that they have to "share them alike," under a Creative Commons license. They will receive full credit for their creation, and the people who will eventually use them will have to do the same.

This modality opens up a virtuous cycle: The knowledge, products, and creations made possible through the Human Ecosystems grow continuously. Anyone who makes something with it contributes to the general benefit of the communities in the cities, who can capture this increased knowledge, do something with it, and share it back into the pool. A virtuous loop is created in which the common resource pool grows continuously.

This is exactly what happened in Sao Paulo, as we will see in the Education section of this chapter.

12.5 Public Presence

The public presence of the DUA process in Sao Paulo was implemented with the support of SESC. As part of its policies, SESC creates outstanding multipurpose centers disseminated in various cities across Brazil. These centers are intended to provide an accessible, usable, and inclusive welfare dedicated to people's culture, intellectual growth, physical well-being, and for the development of their imagination and of their opportunities for socialization and creativity. In SESCs all over the country, it is possible to visit art exhibits; access the library and the spaces of the locations where there are magazines and newspapers from all over the world; meet people and play chess together; eat wonderful organic food at the lowest prices in the cafeterias and bars; rehearse with our band; try out a theater play in a fully equipped venue; access workshops and courses on printmaking, design, ceramics, and languages informatics; use the swimming pool; 3D print something; and more. They are centers for civic life, where the inhabitants of the city can come together, do things, learn others, and understand their city and its life.

All at accessible prices, or even for free or through cheap monthly subscriptions.

The structures themselves are outstanding, featuring urban interventions from architects such as Lina Bo Bardi and her SESC Pompéia, as well as many others.

The SESC Vila Mariana was dedicated to the Human Ecosystems and to its DUA process.

A series of interactive experiences, mainly built around large scale information visualizations, were designed and installed throughout the space of the SESC Vila Mariana, constituting the Real-Time Museum of the city.

Visitors could explore the Relational Ecosystem, the topics and emotions expressed in the city, their geographical and temporal manifestations.

The Real-Time Museum of the city of São Paulo is dedicated to creating an iconic, engaging, and inclusive experience of the Relational Ecosystem of the city. It has been designed to create awareness about its many layers, as well as to provide a cultural shock to the population, to promote a radical transformation of the sense of possibility about the perception of what a metropolis can be and about the ways in which the concept of citizenship can be transformed. Here citizens can access a learning process—first in a facilitated way, through the interactive experiences, then through the education process—to become positive, active, aware, and participatory agents of the city's civic life. The museum hosts a laboratory in which all members of society can learn how to use the Relational Ecosystem of the city, using Digital Urban Acupuncture techniques, to pursue their goals and to create patterns of self-organization and intervention on the urban environment, supported by the city, regional and federal administrations.

12.6 Education

The education phase of the DUA process was built around four main pillars:

- formal training at the university;
- extemporaneous interventions;
- workshops and courses;
- peer-to-peer learning.

The first ones included formal courses and workshops which were given out to researchers and students at the university. We started from our partner, the Universidade Metodista. The courses include basic and advanced notions, dedicated to different types of subjects: from the basics of the methodology, to the ways in which data could be analyzed, to how to create anything from artworks, services, objects and processes based on the Human Ecosystems, and more. These occasions were also useful for acquiring an informed user base who could, on the one hand, help us in the museum (for anything from becoming an assistant for visitors, to the management and maintenance of the systems in the museums, or to updating the exhibits) and, on the other hand, to bootstrap the peer-to-peer learning processes, which we will describe in the "performance" section of this chapter.

The second ones were designed through careful organization of the museum's environment. A series of visual cues and support materials—under the form of signage, brochures, and informative materials—were systematically disseminated in the space of Vila Mariana, so that as many aspects of the overall design would be clear and accessible. People would know what each visualization and interactive experience was about, how to use it, and also get some hints and practical examples of things they could look for or questions which could find answers by using the artworks.

On top of that, trained assistants were constantly available throughout the day to support visitors who would access the visualizations. They would be able to provide general, accessible explanations about the project and the interactive experiences, and also to guide visitors through them, in order to facilitate people's learning curves and understandings. The assistants would also assist visitors in other, more active, ways, for example by engaging conversations which explored what curiosities and questions people would express about their city, and helping them out to find solutions by using the Human Ecosystem, using the visualizations to search through the Relational Ecosystem and the topics and emotional landscape of the city. Whether it was about exploring which expressions people in the city expressed about a certain topic, how they formed communities, how communication and news spread through the city, or about the variation in time of the informational and communicational environment in Sao Paulo, the assistants would suggest ways in which visitors could use the system and experiences to find and understand what they were looking for.

Each session could last a few minutes but could also go up to half an hour or more, as people got involved and wanted to know more. Sometimes groups of people formed, gathering around to benefit from the explanation being given to a single visitor, and grabbing the opportunity to add their own questions and doubts to the conversation.

When this happened, discussions would be transformed into mini public lectures, which would provide further information and also cause rise in curiosity and desire to learn more.

In all of these discussions, whether with single individuals or with groups of them, the critical issues connected with the Human Ecosystems (and, thus, with the Ubiquitous Common) were also explored. Assistants were trained to show how people's relations could be explored, and their microhistories made of locations, emotional expressions, and topics of interest. The critical implications deriving from these possibilities were discussed, confronting with issues such as privacy, the transformation of public, private, and intimate spaces, digital divides and illiteracy, and about the possibilities, on the one hand, for people to learn and to protect themselves and, on the other hand, to understand how this information could be used for positive, constructive reasons, in collaborative ways among citizens of all sorts, promoting a new idea of the city.

People were systematically invited to join in the workshops and courses given out about the Human Ecosystems in Sao Paulo.

Many times it was not clear what the courses were for, what their practical output would be.

While the experience of the interactive information visualizations was almost constantly perfectly clear—maybe after some explaining, especially dedicated to the less digitally literate—the idea that a workshop or a course would help people to embrace new kinds of action in the city was not always clear. Of course, different types of subjects expressed different reactions. Most artists and designers, for example, immediately expressed interest, grasping the opportunity to create new toys, artworks, applications, and services. Architects and engineers would see the opportunities for planning and for the creation of specific projects and digital dashboards to monitor and observe the city and to gain better understandings of citizens' behaviors. Researchers, managers, and policy makers would see the opportunities for gathering insights and strategic expertise.

What was initially missing was the possibility for a wide understanding "on the ground," for people. Many people would establish parallels with the many applications which allow them to visualize their communities and connections on social networks.[2] But they would only rarely understand directly the opportunities coming from the possibility to be able to interact with the Relational Ecosystem of an entire city. This was almost always the case, for example, with elderly people and with children. While the former seemed many times to be afraid or even in open opposition to such a possibility, the latter desired a playful approach, a videogame of the Relational Ecosystem of the city.

In general, multiple questions would arise. "What will I be able to do with this?" "How will it change my life as a citizen?" "How could I make money with it, or create a profession out of it?" "They are my friends on social networks! I can send them a message!"—This, for example, was a constant, because people did not understand that they could explore the entire city, not only their own social network friends.

These were only just a few of the questions which were asked. We did not include any of the ones which referred to technologies and to the fact that many people seemed to feel unfit to access and use such "complex systems," even when they saw them under the form of accessible, simple, usable, and entertaining information visualizations. The transition from spectatorship to active usage was a difficult one. While they were entertained, sincerely interested and engaged by the visualizations and experiences, the idea of learning how to "do things" with it was off putting for many.

We confronted with this kind of issue in three different ways:

- metaphors;
- critical engagement;
- create a sense of objective, or goal.

Through metaphors we were able to break into people's imagination, and to engage them in more significant ways, so that they would actually become interested in acquiring a more active role. For example, one metaphor which seemed to obtain immediate understanding was the one of the forest.

"*If you are a tree in the middle of the forest, you can only see the trees around you, but you cannot see the whole forest. Seeing the trees around you (your friends,*

[2]Like Intel's "Museum of Me" http://museumofme.intel.com/.

the people you know in your neighborhood, your closest connections, people in your school or office) may seem enough already. And sometimes it is. But not being able to see the entire forest, from above, does not allow you to understand the big picture. To see it from above means being able to see if the forest is burning from a side, or if other trees far away are catching a disease which is spreading, or if people are cutting off too many trees, or other things like that. Being able to observe it, from above, allows you to take action, by coordinating with other trees (people) and by bringing their collective voice to the institutions, to stop the fire, halt the spreading of the disease, change the tree-cutting policy. Seeing the overall picture also enables you to understand what trees to interact with in this organization, because this certain tree is the largest in the whole forest, or this other one is more resistant to the disease, or because another one is right near the house of the ranger who can be alerted to stop the fire."

When metaphors were switched over to establish the parallel to the citizens and their behaviors, drastic changes occurred, and most people started understanding the possibilities. Cellular metaphors were also very useful in that, by interpreting the city as a body, and the various types of cells being organized in organs, connected through the nervous system, and so on, which would allow to establish parallels with communities, roles in the Relational Ecosystem and with all of the elements which have been explored in the previous chapters.

The second way in which we would engage people to take active part in the workshops and courses was to suggest critical issues which they would engage on.

Possibly, the most successful one in acquiring people's attention was related to the idea of surveillance. It went like that: "Major organizations have this information about people's behavior (operators, companies, governments, and secret services). They use it to sell us things, to observe and control us. The only ones who don't have information and knowledge about people's behavior are people themselves, who cannot use it to become better citizens, to organize themselves autonomously, to protect themselves, to create new jobs, opportunities, artworks, services and more."

This motivation was very successful in engaging people. It created a sense of urgency and necessity by providing a simple truth: we, as citizens of the world which is progressively becoming more digital and connected, constantly produce—whether we realize it or not—data and information; and we, the citizens, are the only ones who cannot directly benefit from it, because its access is currently reserved for major operators who use it for marketing, social control, political campaigns, and security. Such motivation proved to be very effective, and people would see the opportunity with new eyes and under a new light.

But it was still not enough. Even if the motivations were strong, many people still failed to see what they could achieve with the tools and methods of Digital Urban Acupuncture, including Human Ecosystems. They could understand the purposes and directions, but transforming them into real, pragmatic actions was out of their reach, most of the time.

This is why we decided to create workshops which expressed the possibility to achieve specific goals and objectives. Not generic "Introduction to the Human Ecosystems" workshops, but specific ones which were able to communicate purpose and meaning, such as "How to make our personal mobility better and more efficient?" and "How to fight crime in our neighborhood?" and "How can we organize ourselves among citizens to clean a park together?" and "How can we invent a new job?" and more.

These were much more effective, so much that different things started happening, opening up the participatory performance phase of the DUA process.

12.7 Participatory Performance

As expected, the educational process bootstrapped the participatory performance phase of the DUA.

As described in the previous chapters, the objective of the education process is to enable people how to do certain things, acquire specific methods, and use a set of tools—technological and non-technological—to start autonomously imagining and executing processes in the city, their own Digital Urban Acupunctures, by using the Relational Ecosystem to perform a peer-to-peer ethnography of the city and, thus, understand how to organize together to bring transformation.

The best possible result of DUA's education process is twofold: On the one hand, people learning things, creating their own and teaching other people how to do things and how to create their own (in what is called peer-to-peer learning); on the other hand, the emergence of spontaneous, unexpected processes.

This is what we interpret as a social acceleration, the progressive chain reaction which happens at mental, psychological, social, and interactive levels when the overall process is started. All of the characteristics of the project, with all of its iconic, physical, educational, and psychological elements contribute to the possibility for this emergence.

This is exactly what happened in Sao Paulo.

One day, in September 2014, while we were doing a workshop, the participants asked if we could change the subject of the workshop to explore how people in the city were dealing with a particular subject: water.

When we asked for the reason behind this request, we learned for the first time about Sao Paulo's water crisis.[3]

In the city, an extended dry season and the ever-increasing demand of water have been leading, during the second semester of 2014, to a draught. At the apex of the water crisis the political elections were taking place, in October, finding the political forces unwilling to confront with this issue, to avoid losing consensus. On top of that, the private water companies—using the crisis at their political advantage

[3]For example, as seen on BBC: http://www.bbc.com/news/world-latin-america-29947965.

to make claims and releasing limited and contradictory information—the pollution of the rivers, the needs of the industry, and the marked inequalities which are present in the area all contributed to a chain reaction, whose effect was the sudden emergence of entire areas of the federal state of Sao Paulo which were practically left with no water.

We had never heard about the crisis before, and we welcomed the opportunity to change the subject of the workshop, both for the importance of the topic and for the fact that people were trying to address an issue which was truly relevant for them, expressing the desire for self determination and expression.

Thus, we set forth in observing the expressions about "water" in the city.

We filtered the content, the locations, and emotional expressions using a series of different modalities: Starting from people dealing directly with the water crisis and arriving at more general expressions on the topic of "water," to try to gain understandings if such a sensible issue would influence people's relationship with water, or to try to understand if some unexpected, unaccounted for phenomena would arise in the city.

We were able to perform these kinds of observations by using the natural language analysis features of the Human Ecosystems which have already been described in the previous chapters.

We were able to observe the city of Sao Paulo during most of the crisis. For example, a visualization of the feeling of anxiety for the water crisis, as expressed on social networks, can be seen in Fig. 12.3.

During the observation, we focused on the possibility to identify the places in which negative emotions (such as anxiety, fear, unrest, sadness, and more) would show up, also differentiating them according to their subtopic. We were particularly interested in understanding if it was possible to determine the areas in which people were actually experiencing the effects of the crisis.

While we were able to identify these emotional expressions in Itu, Gaurulhos, and São Mateus (among others), obtaining insights about the draconian measures which were put in place, and their effects on citizens of these neighborhoods, we also understood how several additional considerations must be addressed.

Among the most interesting of these is the possibility of understanding how representative and inclusive this data is.

For this, we confronted our findings with the statistics and geographical distributions of internet access and social media usage in Brazil. With 53 % of the population on the Internet, and 45 % active on social networks,[4] the situation can be described as partially representative and inclusive, also taking into account the disparities in the distribution of wealth, divides and literacy, both in absolute terms and in relative ones, across geography, status, cultures.

[4]Source: ComScore MediaMetrix: Brazil, Digital Future in Focus, Feb. 2014, available at: https://www.comscore.com/Insights/Presentations-and-Whitepapers/2014/2014-Brazil-Digital-Future-in-Focus-Webinar.

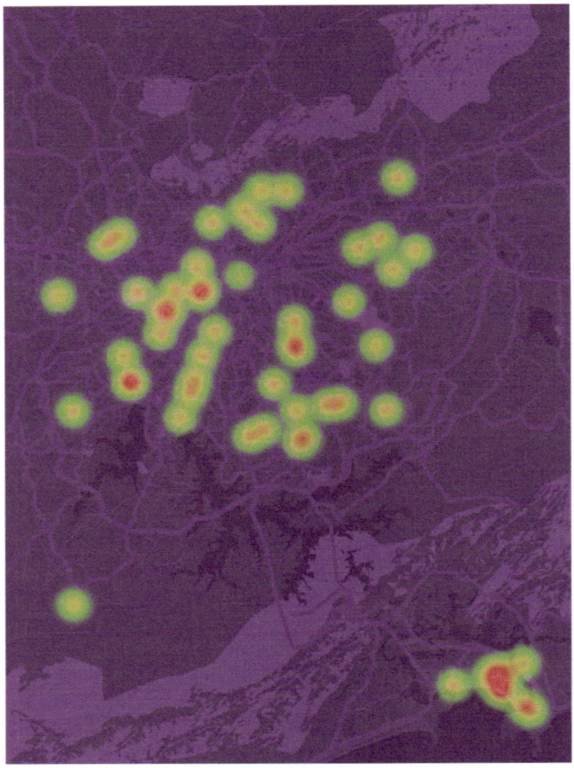

Fig. 12.3 The geographical distributions of the emotions of anxiety, as expressed on social networks, about the water crisis in Sao Paulo, from the beginning of April to the end of November 2014, showing 387,173 user-generated contents

To address this issue, processes and actions should be designed to reach substantial parts of the remaining part of the population.

Some of the other issues which we have encountered are technical in nature.

For example, the possibility to deal with irony using the algorithmic approaches typical of natural language processing. Irony and humor have an extraordinary presence on social media. While there are partially successful approaches in dealing with these expressive modalities—as shown in Bermingham and Smeaton (2010) and in Carvalho et al. (2009), among the others—this remains an open issue and can be mitigated by adopting multiple approaches and by analyzing content in context.

We worked directly with the workshop participants well beyond the duration of what had been intended to be a one-day initiative, extending the process for almost two months and, then, giving them full access to the platforms to keep the observation going.

The findings were exceptional.

In regard to the emotional analysis (whose statistics are shown in Fig. 12.4), anger was the most widely expressed feeling, focused on corruption, the

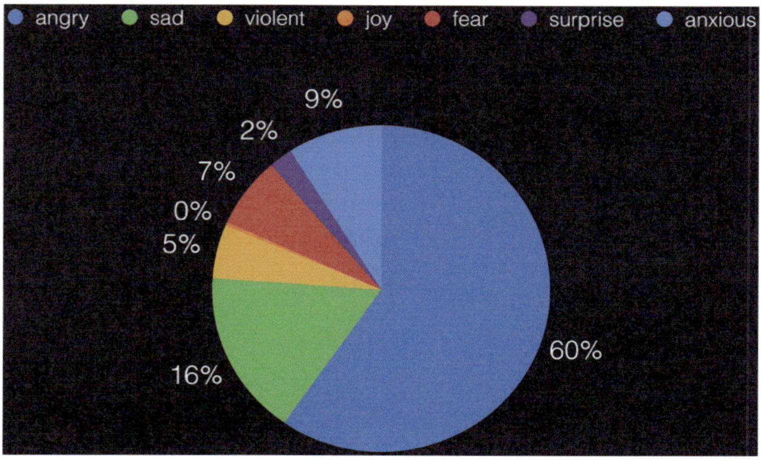

Fig. 12.4 Emotions about the water crisis in Sao Paulo

malpractices of the government, and on the misbehaviors of industries and industrial agriculture. Anxiety and sadness followed immediately behind, with wide areas of the population expressing these feelings in regard to the difficulties posed by the situation onto their daily lives, health, and security (e.g., for hospitals and emergencies).

Violence and fear were also present, the second often happening as a result of the first, as many people expressed about possible riots and uprisings.

A small part of the expressions (about 0.2 %) were about joy, as they manifested how this difficult scenario had provoked a heightened sense of community and solidarity.

As described in Fig. 12.5, the main topics discussed when dealing with the water crisis regarded the hypotheses (or certainties) of corruption and the attribution of misbehaviors to the government at all levels (from local to federal). Other fundamental issue was represented by the impacts of the water crisis on people's daily lives: from not being able to taking a shower in the warm Brazilian spring and summer, to the effects of lack of water in homes, restaurants, and up to hospitals and other emergency scenarios which took place in the city, including crime. Of notable importance is the description of privilege: about 3 % of the people referred systematically on how a privileged part of the population did not suffer the crisis like the rest of the population.

Other important focus was constituted by the identification of the "places of the crisis," those iconic or particularly significant locations which were systematically associated with the water crisis in people's expressions and imaginations.

One of them was La Cantareira, Sao Paulo's main water reserve. This is an important location in the city, as it is a large water surface which goes well beyond

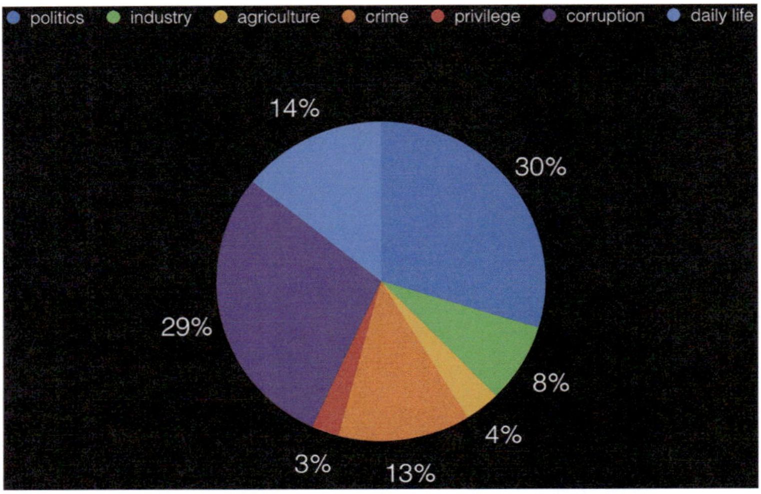

Fig. 12.5 Topics discussed in the water crisis in Sao Paulo

the fact that it provides water for a large part of the city: It is a major landmark for transportation and mobility, as it has large, high bridges crossing from one side to the other; and it is also a place for urban life, as citizens go there for picnics and to enjoy the Sun. La Cantareira was one of the protagonists of people's expressions during the crisis. Both the texts and the images mentioning it were strong and powerful, and it seemed to embody the troubles of the city. To all effects, using all of the images which were shot by Sao Paulo's citizens during the crisis, it was possible to monitor the level of water reserves, by simple visual comparison, as shown in Fig. 12.6.

Multiple interfaces were designed to observe the people of the water crisis, as they expressed on social networks while it happened. One of them is shown in Fig. 12.7, highlighting people's social profile images as they expressed in real time, as well as their connections in the Relational Ecosystem. Such tools were useful and beautiful, as they conveyed the complexity of the phenomenon.

Dealing with the images captured from social networks, it was possible to gain a visual, clear understanding of people's behaviors in the city. Scenes of protests played a major role in this aspect. But they were not the only ones.

Many images, as described by the collage in Fig. 12.8, showed a multiplicity of different behaviors, such as the ones described in the image above. People inventing themselves new jobs through the water crisis, delivering water in various ways. People started to think together, gaining new sense of family and of collaboration. Other inhabitants inventing ways to water their plants or to generate water. Citizens embracing artistic, poetic actions, like the mother and child who took series of

Fig. 12.6 The places of the water crisis in Sao Paulo: La Cantareira

Fig. 12.7 People of the water crisis in Sao Paulo

pictures in front of the water reserves before, during and after the water crisis. Even marriages due to the crisis were documented: As people could not cook, shower, and perform other basic tasks for the lack of water, they started collaborating and doing things together; so much that some of them married and posted images of it on major social networks, explicitly correlating them to the water crisis. There were also critical, potentially illegal and dangerous activities which were being documented: For example, one person published on Facebook a complete tutorial about how to recognize from the others a specific type of manhole on Sao Paulo's streets, which is used to conduct emergency water supplies (e.g., for hospitals); the tutorial was a success, so much that multiple images can be seen about people using these manholes to illicitly draw water.

Specific Web sites were built using this data and information. For example, the "Tem Agua Na Cantareira"[5] which showed through a half empty/half full glass the level of water contained in the reserves, as shown in Fig. 12.9.

It was an incredible cultural richness with myriads of possible usages, narrated through the texts, images, and expressions coming from the daily lives of the inhabitants of Sao Paulo.

This wealth, collected, became the first official collaborative production of the Human Ecosystems in Sao Paulo. It was a truly engaging experience, with multiple

[5]"Is there water in La Cantareira?" http://temaguanacantareira.com.

Fig. 12.8 A collage of people's behaviors during the water crisis in Sao Paulo

citizens understanding a new concept of citizenship, while they were observing the city, collecting and understanding its expressions, engaging it and its inhabitants through social networks, and creating shared, practical actions, such as protests, software, image collections, analyses, and research.

At the time of writing, the process is still ongoing, and it is planned to perform a nationwide exhibit together with SESC, featured in 29 locations across Brazil.

This was only one of the acts of spontaneous participation which were triggered by the DUA process in Sao Paulo. The many more which were about to come led to the conclusions that the opportunity was present to engage citizens across cultures into a radical reconsideration of what citizenship could be, and on how to move from a governance based on consensus to one based on the possibility of co-existence of multiple active points of view in the city, and on the enormous richness which this fact represents.

This is what we will deal with in the next chapter.

Fig. 12.9 Tem Agua Na Cantareira

References

Berlyne DE (1971) Aesthetics and psychobiology. Appleton-Century-Crofts, New York

Bermingham A, Smeaton A (2010) Classifying sentiment in microblogs: is brevity an advantage? CIKM'10, October 26–29, Toronto

Carvalho P, Sarmento L, Silva MJ, de Oliveira E (2009) Clues for detecting irony in user-generated contents: oh…!! it's "so easy";-). In: TSA '09 proceedings of the 1st international CIKM workshop on topic-sentiment analysis for mass opinion, pp 53–56

City of Vilnius (2013) The happy barometer. http://happybarometer.com/. Accessed 12 Nov 2015

Davidson J, Milligan C (2004) Embodying emotion sensing space: introducing emotional geographies. Soc Cult Geogr 5(4):523–532

Desmet PMA (2002) Designing emotions. ISBN: 90-9015877-4

Frijda NH (1986) The emotions. Cambridge University Press, Cambridge

Iaconesi S, Persico O (2011) Consciousness of streams. http://cos.artisopensource.net/. Accessed 12 Nov 2015

Iaconesi S, Persico O (2014) Urban acupuncture in the Era of Ubiquitous Media. J Community Inf 10(3)

Kaplan HA (1987) The psychopathology of nostalgia. Psychoanal Rev 74:465–486

Lazarus RS (1991) Emotion and adaptation. Oxford University Press, Oxford

London School of Economics (2011) Mappiness. http://www.mappiness.org.uk/. Accessed 12 Nov 2015

Maslow AH (1943) A theory of human motivation. Psychol Rev 50(4):370–396

Nold C (2004) Biomapping. http://biomapping.net/. Accessed 12 Nov 2015

Nold C (2007) Emotional cartography. http://emotionalcartography.net/. Accessed 12 Nov 2015

Ortony A, Clore GL, Collins A (1988) The cognitive structure of emotions. Cambridge University Press, Cambridge

Purcell AT (1986) A schema discrepancy model of environmental perception and affect. Environ Behav 18:3–30

Roseman IJ, Smith CA, Scherer K (ed), Schorr A (ed), Johnstone T (ed) (2001) Appraisal theory. Appraisal processes in emotion: theory, methods, research. Oxford University Press, Oxford

Russell JA (1980) A circumplex model of affect. J Pers Soc Psychol 39(6):1161–1178

Ulrich RS, Altman I, Wohlwill JF (eds) (1983) Aesthetic and affective response to natural environment. Behavior and the natural environment. Plenum Press, New York

Vigneshwara M (2013) Aleph of emotions. http://www.mithru.com/projects/aleph.html. Accessed 12 Nov 2015

von Bismarck J, Maus B, Wilhelmer R (2008) Fuehlometer. http://www.feld.is/projects/fuehlometer/. Accessed 12 Nov 2015

Zajonc RB (1980) Feeling and thinking: preferences need no inferences. Am Psychol 35 (2):151–175

Chapter 13
Ecosystems: From Consensus to Coexistence

Abstract This chapter sums up all the observations narrated in the previous sections to describe a series of philosophical and methodological results which are central to the Human/Relational Ecosystem and Digital Urban Acupuncture approaches. These can best be synthesized as a shift in focus: away from the idea of the necessity of consensus and toward the idea of the possibility and feasibility for coexistence, valuing differences, and turning them into the creative energy driving the well-being of the city, establishing a parallel with the bio-diversity of natural environments.

In this chapter we will sum up all the observations narrated in the previous ones, to describe a series of philosophical and methodological results which are central to the Human/Relational Ecosystem and Digital Urban Acupuncture approaches. These can best be synthesized as a shift in paradigm. Moving away from an idea of governance (of the city, but it is worthwhile reflecting on this issue also from wider perspectives) which is based on the capability to form consensus, to one which is able to interweave different perspectives in dynamic, ever-changing, syncretic equilibriums, defining the possibility for the coexistence of multiple orientations, valuing differences, and turning them into the creative energy driving the well-being of the city, establishing a parallel with the bio-diversity of natural environments.

This shift opens up a novel role for institutions and organizations: from decision-makers to enzymes, facilitators, enablers, and contributors to the dynamic preservation of commons.

In this reflection, we may start from a question: "Where is the city? Where are its cores and boundaries? … How can we grasp what a city is and could be?" (Healey 2002).

Patsy Healey considers this question in a fractal way (Ultramari and Firmini 2010), tracking the diverse interpretations coming from the actors who live, build, and use cities. She describes "imaginative," socially constructed concepts of city, and those that are "locked into cultural inheritances and institutional practices" (Healey 2002).

S. Iaconesi and O. Persico, *Digital Urban Acupuncture*
DOI 10.1007/978-3-319-43403-2_13

"*Cities are not just material artefacts, although we experience the materiality of urban life [...] what is city lies beyond these specific existences and materialities, although as conceived, cities are full of people, technologies and power relations. But yet, cities do exist and have a material effects*" (Healey 2002).

This vision may be explicited to include the informational landscape of the city, arriving at Marcos Novak's definition of "liquid architecture" and "transarchitecture" to explain the complex relations between the physical space and networks of interactions in cyberspace. As expressed by Mike Crang:

"*The architect Marcos Novak offers one way of thinking through these issues in his projects to create, first, a liquid architecture of cyberspace, which he suggests offers an 'augmented space', that is thinking through what worlds of information might be shaped like, and, second, a 'transarchitecture' of their intersection with material world [...] Novak suggests that both mean redefining the urban field by challenging three deeply embedded assumptions of urban studies. First, that space is three-dimensional and shared between actors. Second, space is either solid or void. And, third, you can only be in one place at one time.*" (Crang 2000).

Crang cites Labbeus Woods as an example of the theoretical and practical approaches which deal with the possibility of addressing parallel cities called "centri-cities," expressed by complex interactions and differences:

"*From the mid-1980s he [Lebbeus Woods] produced the idea of 'centri-cities', formed of overlapping interference wave patterns expressing life in a multipolar urban city. Urban multiplicity stands in opposition to the classical city—where the acropolis represented the single centre of authority that worked hierarchically through the polis [...] Instead of utopian monologue producing the hierarchical city (organized around the one principle) he looks for a heterarchical city of dialogue that is necessarily incomplete and incoherent.*" (Crang 2000).

These notions make it impossible to define frontiers and boundaries for the contemporary city, for example, as indicated by Skeates (1997) who describes a new type of urbanity in which the city and the countryside come together.

This is where the opportunities of technologies are staged, allowing global communication, overcoming frictions of distance for economic transactions and human interactions, and challenging perception and comprehension of space and time and, thus, of the city itself. According to Skeates, territorial and spatial organizations are under threat, and this also corresponds to a basic misuse of the terms which refer to space and time.

"*We are beginning to understand that there has been a shift, a break with the past that means that we can no longer use the term 'city' in the way that it has been used to describe an entity which, however big and bloated, is still recognisable as a limited and bounded structure which occupies a specific space.*" (Skeates 1997).

This conceptualization of the city also gives rise to the notion of "artificialized nature," as everything becomes reachable and interconnectable through technologies. This is very similar to Santos' (1994) conception of the "technical scientific-informational milieu," where technology, science, and information are totally embedded in space and spread across the territory, "artificializing" nature.

"The so-called postmodern city represents a huge leap away from any referent that can anchor that meaning outside of itself. All space succumbs to the logic of the constructed, the artificial and the mythologized which attempts to deny the possibility of there being any 'real' conditions of urban existence underneath the layers of hyperreality." (Skeates 1997).

If traditional architecture was place-bound and strongly connected to a certain experience or condition, now mediated environments challenge time and space and architecture can no longer be bound by their supposed static conditions.

In "Inscribing a Square," Offenhuber (2012) recalls how in Alpaville (1965) director Jean-Luc Godard leads the viewers through a journey in a city controlled by an omniscient computer "Alpha 60." The movie's originally intended title, "Tarzan versus IBM" (Darke 2005) would have been a better match for today's main interpretation of the evolution of cities with technologies: on the one hand, the ubiquitous arrival of smart city solutions promoted by companies such as IBM and CISCO in order to upgrade and optimize their urban infrastructure; on the other hand, the rise and success of civic protests and disobedience coordinated from the bottom-up, such as the Arab Spring, Wikileaks, the Spanish May 15 h, Occupy and more. Offenhuber's book describes how information shapes the city: "its sensory experience, its infrastructures, and its places. We are interested in the ways different groups use urban information to make sense of public spaces and change them."

The actions of "sensing place" and "placing sense" become the mutual feedback loop which seals the description of the city in terms of a multifaceted entity endowed with a networked nervous system.

13.1 The City as a Body

In his "Sentient City Survival Kit" Shepard (2009) declares:

"The Sentient City thus becomes a contested site: a theoretical construct within which longstanding claims of essential human qualities, capabilities and characteristics are critically destabilized through their attribution to non-human actors. This destabilization is understood to work actively, as a tactical maneuver enabling other ways of thinking that not so much confront dominant ideologies but elide common wisdoms about not only what it means to be human but also what it might mean to be a city. In gathering archaeological 'evidence' of near future urban conditions, the Survival Kit enters the debate on non-human sentience through the trapdoor in the floor."

This modality (the "trapdoor") is indicated through a series of comparisons. Among the others, Shepard uses MIT's Serendipity project,[1] drawing on the real-time sensing of proximate others using Bluetooth technologies built into mobile phones to search for matching patterns in profiles of people's interests:

---

[1]Serendipity. http://reality.media.mit.edu/serendipity.php.

"Developed by the Human Dynamics Group at the Media Lab, the project's goal is to facilitate corporate productivity by providing a matchmaking service for workers with shared interests or complementary needs and skills who otherwise might not encounter each other within spaces organized around the office cubicle."

Eagle (2004) describes multiple possible scenarios, mostly dealing with people in the same office discovering each other through their network-connected devices, placed in various versions of "available" modes, vibrating or signaling in other forms whenever a fellow coworker would be dealing with (or having the expertise in) some tasks that the first office mate would be interested in. These would create a highlight on a certain common interest and, in the project's concept, it would also create the value deriving from the fact that the two fellow coworkers would not have stopped to talk, otherwise, thus missing an opportunity.

Shepard (2009) notices how the term "matchmaking" used in the project is by no means isomorphic to "serendipity," when understood in its common definition of "looking for something else." In effect, it is the exact opposite: "it simply outsources the problem of finding something we are already looking for." On top of that the author criticizes the introduction of the "available" modes, suggesting some attempt to address privacy issues with little consideration in making transparent who can access and use your profile data.

In response, Shepard describes his Serendipitor project:

"*The GPS Serendipitor is an alternative GPS navigation software application for mobile phones that determines a route to a destination that the user has not previously taken, designed to facilitate finding something by looking for something else. What are the implications of a society that needs to download an application for serendipity?*"

This tension (and anxiety) in the definition of the "smart" components of our current and future cities constitutes one of the main conflicts in today's urban scenarios.

This is also evident, for example, on Espen Aarseth's "cyborg aesthetic" debate (1997).

Donna Harroway's Cyborg Manifesto of 1991 used the cyborg, "a hybrid of machine and organism," to challenge the notions of gender, nature, race, and identity. Aarseth speculates on the cyborg as a means to problematize power and control structures (Khan 2011). He writes,

"*Any cyborg field, as any communicative field, is dominated by the issue of domination and control. The key question in cyborg aesthetics is therefore, who or what controls the text? Ideologically there are three positions in this struggle: author control, text control and reader control.*"

In the book, Aarseth used John Cayley's Book Unbound software program from 1995 to show an algorithmic interpretation of human interaction, highlighting it as a novel form of cyborg and, thus, as a yet unexplored version of identity, as expressed by merging human expression and machine interpretation: "The text is an impurity, a site of struggle between medium, sign, and operator. The fragments produced are clearly not authored by anyone. They are pulverized and reconnected echoes of meaning, and the meaning that can be made from them is not the meaning that once existed [...] The pleasure of this text is far from accidental; it belongs not to the illusion of control but to the suggestive reality of unique and unrepeatable signification."

How can these observations become significant when observing the evolution of cities? A possible hint may originate from the observation of technologically mediated environments—as our cities—where ubiquitous technologies constantly draw data and information from inhabitants to be processed through algorithms, to produce other data and information which is fed back to the people under a variety of different forms, thus influencing their behaviors. A cybernetic system forms.

On top of that, this type of cybernetic system is not related to a single individual, but can be implemented by algorithmically recombining the data and information of, potentially, the entire population of the city.

Enter the cyborg city.

13.2 Cybernetic Organizations

Paul Dourish's work on embodied interactions from 2004 recognizes "that action and meaning arise in specific settings—physical, social, organizational, cultural, and so forth," and that "meaning is conveyed not simply through digital encodings, but through the way that computation enlivens those encodings with semantic and effective power." This thought expresses the fact that our lived context constitutes an integral component in human–computer interactions.

In turn, this also means including a wider set of interacting parts in the design of interactive systems: people, places, and things, in a feedback loop with one another. This would require a profound reflection on the architecture of cybernetic systems and about the ways in which they can be implemented in the scenarios of our everyday lives. This is the kind of effort which has been performed in works such as "What is Interaction? Are there different types?" by Dubberly et al. (2009), in which they propose an expanded taxonomy of interaction. This kind of work is interesting in its intention in describing interactive systems as a set of relations which are linear, self-regulating, and learning, that combined with one another manage to create complex interactions and systems which apprehend and mutate, exploring behaviors from a systems design approach.

The distinction between linear, self-regulating and learning systems is not casual, as it comes out from the history of cybernetics (Weiner 1948).

Linear systems take an input and produce some predictable output, demonstrating clear cause and effect relationships: They are reactive, more than interactive. Self-regulating systems have specific goals which are pursued cyclically through feedback relationships with one or more interconnected subjects. In cybernetic terms, control is distributed among all constituents in these relationships, since one influences the others. Learning systems go beyond that, in the fact that they are able to evaluate their goals and change them in order to follow new ones. This is the way in which sentient systems work, learning from their contexts to change not only their actions, but also their directions. Learning systems are even more shifted toward being relational systems, in the fact that not only behaviors change according to the context and its participants, but also the strategies used to achieve them.

In all of this, consideration can be given to the fact that, through ubiquitous technologies, shifts in agency can happen, affecting the biological, technological, and social systems which are parts of the classic cybernetic approach (Foerster 1952).

With developments such as the Internet of Things, we may face the situation in which our devices, buildings, objects, and cities may require less or no involvement from us to perform their functionalities: Devices and objects may initiate communication autonomously, and autonomously carry out tasks: "From anytime, anyplace connectivity for anyone, we will now have connectivity for anything" (International Telecommunications Union 2005). Digitally enabled things will autonomously produce, sense, and process information locally and share it globally with other devices and human beings.

In this case, human participation need not be a central part of these processes.

One way to think about it is "automation," where machines are tasked with performing work in our place.

Another way to think about this is in terms of "interpassivity," the "uncanny double" of interactivity defined by Slavoj Zizek as "the obverse of interacting with the object (instead of just passively following the show) is the situation in which the object itself takes from me, deprives me of, my own passivity, so that it is the object itself which enjoys the show instead of me, relieving me of the duty to enjoy myself" (Zizek 2007). We may interpret it, thus, as a shift in agency.

In yet different terms, we could describe this scenario in terms of the formation of a wider, more inclusive, interconnective nervous system.

In his interview with William J. Mitchell, Harkin (2005) says: "Mitchell's theory is that the city has always been molding us into technology-dependent cyborgs, but that the new communications technologies have made all this more vivid by overlaying on the urban landscape a kind of central nervous system that plugs us deep into the wireless ether." And, then, continues reporting Mitchell's opinion when they come across a youngster deeply immersed in the music coming from his portable device:

"*but mobile devices can also increase your awareness of what is going on around you - people exchanging digital content, for example. Young people do it wonderfully, he says. They can switch attention, multitask, be involved in a real conversation at the same time as they surf the internet and send messages on their mobiles. But are they not not in danger of becoming permanently distracted? Not at all, says Mitchell. In his graduate classes at MIT, all his students bring in wireless laptops, and many of them check Google to verify and enrich what he's telling them. The result, he claims, makes lessons much more open-ended and dynamic.*"

The discourse on "cyborg urbanization" is emerging through the technological, scientific, and medical advances that have transformed everyday life in our cities.

The original "cyborg" definition by Clynes and Kline alluded to the possibility for cyborgs to function in hostile environments and to become extensions of conscious and unconscious human functions, beyond physical limitations and also in terms of expanding human creative potential. Dealing with cities, the urban environment itself may be described as a cyborg linking human beings, objects, plants, buildings, and other things through complex networks, as well as the interaction between technologies and individual bodies. Both interpretations leave open doubts in regard to the interactions with society, for example, for all that concerns the relationships between technology and the exercise of power.

When the ideas of the cyborg city coming from enthusiasts such as Mitchell, Suzuki, and others are placed in a historical context, they face the questions of governamentality in a Foucauldian sense, and the disciplinary and encoding effects of technologies on cultures and diversities.

Nonetheless, the figure of the cyborg has been useful in analyzing the transformations in the relations between nature and culture and in the description of the new political economies which arise in the production of the informational space and of the hybrid forms of identities which arise in these times.

The cyborg is both the body–technology hybrid in the contemporary city and the adjustments to those approaches which desire to mark the supremacy of the virtual or of the physical. It can "rematerialize" the city and establish new forms of relation in the process, creating the chance to build new conceptual vocabularies with which to describe the mutating relationship between body and technology in the contemporary city.

Thus, the city can be described in terms of a cybernetic organization. This, and the possibility and opportunity to go beyond this concept.

For example, drawing from the arts, and in particular those arts which have explored further types of convergences and synergies.

In 2000 Roy Ascott wrote:

"*Between the dry world of virtuality and the wet world of biology lies a moist domain, a new interspace of potentiality and promise. I want to suggest that Moistmedia (comprising bits, atoms, neurons, and genes) will constitute the substrate of the art of our new century, a transformative art concerned with the construction of a fluid reality. This will mean the spread of intelligence to every part of the built environment coupled with recognition of the intelligence that lies within every part of the living planet. This burgeoning awareness is technoetic: techne and gnosis combined into a new knowledge of the world, a connective mind that is spawning new realities and new definitions of life and human identity. This mind will in turn seek new forms of embodiment and of articulation.*"

Just as we have seen in previous chapters, researchers such as de Kerckhove and Gabrys and artists such as Ascott have been exploring new domains, in which psychic space and cyberspace become connected and mutually and fluidly influencing each other. From Gabry's telepathic urbanism, to de Kerckhove's psychotechnologies, to Ascott's moistmedia, life stops being a matter of differentiating between human beings, objects, buildings, and "nature" and engages the possibility for fluid passages from one domain to the other.

Contemporary technologies go in this direction, as our minds, bodies, and behaviors are entangled in networks of physical and digital interactions (so much that it progressively stops making sense differentiating the one from the other), just as objects, places, spaces, and times start gaining the ability to become interconnected into these networks, opening up novel possibilities for communication, interaction, and the generation of sense, intelligence, and identity.

In his "Fourth Revolution" Luciano Floridi (2014) describes how "prehistory and history work like adverbs: They tell us how people live, not when or where they live." In this sense, current human life stretches across three ages: prehistory, or without recorded documents; history, in which societies use ICT to record, transmit, and use data of all kinds; and hyperhistory, "in societies and environments where ICTs and their data-processing capabilities are not just important but essential conditions for the maintenance and any further development of societal welfare, personal well-being, and overall flourishing." In hyperhistoric times, exponentially rising computing power becomes available at exponentially diminishing costs. This computational power ends up in machine-to-machine, human-to-machine, and mediated human-to-X interactions, where X can currently assume the form of other human beings, plants, buildings, software agents, entire companies or governments, and a series of different types of identities, from massively collective ones to anonymous ones, to all there is in-between, including ones for which their humanity/non-humanity/otherness cannot be determined (as we have seen in Chap. 11).

As we have seen in Chaps. 4 and 11, this scenario perfectly fits the ones currently found in our cities. For example, in the Ubiquitous Common approach, used in Digital Urban Acupuncture practices, actors of the ecosystem can assume different connotations:

- *anonymous*: a participant to the P2P network whose identity may be undisclosed for particular reasons;
- *individual*: a participant whose identity is associated with the one of a certain, single, legal person (e.g.: John Smith, or ABC Ltd.);
- *collective*: a participant whose identity is associated with a concept describing a set of subjects (e.g. farmers, citizens of town X, the people associated with the Y association) or a goal (e.g. civic action, ethical scientific research, ancient seeds cultivation);
- *nomadic*: a participant whose identity is associated with a shifting set of legal persons, one at the time, for particular reasons (e.g. first it is John Smith, then John Smith passes it on to Mark White, then Mark White passes it on to ABC ltd.);
- *temporary*: a participant (of the anonymous, individual, collective or nomadic types) whose identity is limited in time, for particular reasons (e.g. for an event, for a project).

All of these actors, in the current state of things, can be embodied in human beings, software agents, plants, buildings, communities, organizations, or even intermixed, cross-species, cross-typology aggregations, all potentially interconnected through ICT.

All of this corresponds to incredible amounts of processing power, most of which is "invisible to us, like the oxygen we breathe, but they are becoming almost as vital, and they are growing exponentially." (Floridi 2014).

This computational power is used continuously process the enormous quantities of data and information which we (all the types of "we") produce, consciously or unconsciously, generating Big Data which can be observed and understood, influencing the entire ecosystem, its behaviors, perceptions, and understanding of the world.

"The real epistemological problem with big data is small patterns. [...] Small patterns matter because, in hyperhistory, they represent the new frontier of innovation and competition, from science to business, from governance to social policies, from security to safety. In a free and open marketplace of ideas, if someone else can exploit the small patterns earlier and more successfully than you do, you might quickly be out of business, miss a fundamental discovery and the corresponding Nobel, or put your country in serious danger." (Floridi 2014).

All these small patterns describe a completely new scenario which is ecosystemic in nature, and which describes the notion of how present and (near) future information ecosystems (and, thus, governance, markets, education, environment, energy, etc.) will not be based on consensus, such as are, for example, our current democratic institutions. Rather they will be based on the idea of coexistence: on the fact that the most value (under whatever form, be it environmental, financial, or related to people's well-being) will be generated by interweaving as many as these small patterns as possible, fitting them into syncretic equilibriums which dynamically change at incredibly high speeds. The possibility to radically recombine the world in as many expressions of diversity and variety will drive future developments, shifting from the logic of consensus to the one of dynamic coexistences.

This process has, of course, serious implications, for example, under the points of view of its accessibility and perceivability; from the point of view of the possibility for autonomy and transgression; and from the point of view of the transformation of identities.

As for the first two: Divides could reach extremes. Either visible and measurable divides (such as digital divides, literacies, cultural divides, and economic divides), or invisible ones, or ones which can only be measured in complex ways (such as the divides which are created through the opacity of algorithms, or through performance issues, as for example the ones which are brought into existence when the "decision-maker" is a software agent making billions of decision each second).

As for the third, the transition to novel forms of identity (and, thus, of responsibility, liability, ethics, and morals) will potentially bring on issues which are social, political, administrative, financial, legal, and psychological.

Is there a solution to these important issues? This currently is an open question, with a possible solution likely to be found in a radical transformation in our current conception of institutions and organizations: from decision-makers and executors to enzymes, and active participants in the preservation of commons and of the high quality Relational Ecosystems which are needed to collaboratively manage them, in inclusive ways. As shown in previous chapters: Institutions should transform into what seems like gardeners of gardens without a form or definite boundary.

References

Ascott R (2000) Edge-life: technoetic structures and moist media. Art, technology, consciousness (HB). Intellect Books, London
Crang M (2000) Urban morphology and the shaping of the transmissable City. City 4(3):303–315
Darke C (2005) Alphaville. French film guides. University of Illinois Press, Champaign
Dourish P (2004) Where the action is—the foundations of embodied interaction. MIT Press, Cambridge
Dubberly H, Pangaro P, Haque U (2009) What is interaction? Are there different types? Interactions v.XVI.1
Eagle N (2004) Can serendipity be planned? MIT Sloan Manage Rev 46(1):10–14
Espen JA (1997) Cybertext—perspectives on ergodic literature. Johns Hopkins University Press, Baltimore
Floridi L (2014) The fourth revolution: how the infosphere is reshaping human reality. Oxford University Press, Oxford
Harkin J (2005) Cyborg city: an interview with William J. Mitchell. http://www.theguardian.com/technology/2005/nov/26/news.comment. Accessed 12 Nov 2015
Healey P (2002) On creating the city as a collective resource. Urban Stud 39(10):1777–1792
International Telecommunications Union (2005) The Internet of Things- Executive Summary. International Telecommunication Union (ITU), Geneva
Khan O, Shepard M (eds) (2011) Interaction anxieties. Sentient City: ubiquitous computing, architecture, and the future of urban space. MIT Press, Cambridge
Offenhuber D (2012) Inscribing a square: urban data as public space. Springer, New York
Santos M (1994) Técnica, Espaço, Tempo: Globalização e Meio Técnico-Científico Informacional. Hucitec, São Paulo
Shepard M (2009) Sentient City Survival Kit: archaeology of the near future. After mobile media, digital arts and culture 2009, Arts Computation Engineering, UC Irvine
Skeates R (1997) The infinite City. City 8:5–20
Ultramari C, Firmino R (2010) Urban beings or city dwellers? The complementary concepts of urban and city. City Time 4(3):29–40
von Foerster H (ed) (1952) Cybernetics, circular causal and feedback mechanisms in biological and social systems: transactions of the Josiah Macy conferences on cybernetics 1950–54. Josiah Macy, Jr. Foundation, New York
Weiner N (1948) Cybernetics: or control and communication in the animal and the machine. MIT Press, Cambridge
Zizek S (2007) How to read lacan. W. W. Norton & Company, New York

Chapter 14
A Temporary Conclusion

Abstract In this chapter, we will temporarily end our journey through Digital Urban Acupuncture by making some remarks about the origins of this research and about its future directions.

14.1 The Present

In the introduction, we started the discourse by highlighting the profound transformations of our time and the conflicts that came (and that are coming) with it.

There are a number of topics on which these conflicts will be arising, and all of the disciplines which we have been dealing with in this book will be profoundly affected by them. Possibly all transformations will revolve around the radical and continuous mutation of the definition of what we perceive to be our public, private, and intimate spheres. As a consequence, we will see transformations in how control and power manifest themselves in our societies, and on the state of surveillance and on the possibility to achieve privacy and respect for our rights.

Another issue will be the one of inclusion and access, and, therefore, of the transformation of divides and literacies. If, on the one hand, all the developments which we are experiencing come as a welcomed opportunity for more inclusive scenarios, on the other hand, it is evident how some parts of society will be able to be more exposed to the benefits which originate from them. This is valid now and is clearly visible from the many void spots on the maps and visualizations which we have been looking at all over the book: as in all of these maps, what is on the map is just as interesting as what is not on it, as it is the result of divides, lack of infrastructures, differences in literacy and access to knowledge, sustainable economies, jobs, and education. This will be true in the future, as well, unless we become more aware of these issues, and get ready to address them.

The opacity of the algorithmic domains and of our possibility to gain better understandings about our interactions with them will constitute another large part of the conflict. This is visible now through all sorts of crises, crashes, bugs, "unexpected" behaviors, security breaches, cyberattacks, and algorithmic faults which we

S. Iaconesi and O. Persico, *Digital Urban Acupuncture*
DOI 10.1007/978-3-319-43403-2_14

have very little control upon, and which will get worse as our lives turn to be more deeply entangled with the posthuman entities.

For this, law and legal systems will be other foci of the conflict. As technologies—and the practices that come with them—accelerate, the law does not seem able to keep up to speed, losing ground and enabling erosion of many of our basic rights, in ways which are explicit and evident, and also completely opaque and hard to discern.

In our opinion, and given all of the previous observations, *conflict* is one topic that deserves careful reflection, and this is among the directions in which we are orienting a large part of the future of our research. What is the role of conflict in the future of human transformation? How is it possible to recognize it as a value, as the evidence of the many differences in the world, which ensure its diversity and, thus, resistance and resilience?

For example, the many different narratives of the smart city currently fail in describing a place for conflict and transgression.

During the High Level Group meeting on Smart Cities, in Brussels, on September 24, 2014, architect Rem Koolhaas[1] has posed a very simple question: "Where is the possibility of transgression [in smart cities]?" Adding, then: "Smart cities and politics have been diverging, growing in separate worlds. It is absolutely critical that the two converge again."

Transgression. This term defines multiple different concepts.

Here, we refer to its sociological understanding, according to which we transgress whenever we infringe a certain social norm. Transgression implies going beyond a boundary, a limit, but also its existence. As described by Bataille (1957) in Eroticism: "Transgression opens the doors towards what is beyond the limits of what is usually observed, preserving them." There is little space for transgression in this age of smart (smart cities, smart communities), of innovation and creativity. Thus, there is little space for conflict.

The creative class has already been absorbed by the industry. Hackers, makers, startuppers, and the other human profiles in the new scenario create the ranks of the new research laboratories and of the production lines of the industrial complex. They are the unexpected blue collars of this type of industry, perfectly encoded in the new models of labor and production. This economy relates to an industry which has understood creative thinking as its prerequisite, as highlighted by Pine and Gilmore in their "Experience Economy" (1999). There is no doubt, from the point of view of the architectures of power, on who runs the show.

Troublemakers, in this scenario, have become precious commodities. Hans Magnus Enzensberger used this exact word in his "Industrialization of the Mind essay" in 1982. According to his thesis, the cultural industry lives in a state of paradox: It cannot produce its own product (conscience), as conscience is a social product and, thus, it can only be induced and reproduced by them. A sterile industry

[1]Available here: http://ec.europa.eu/archives/commission_2010-2014/kroes/en/content/my-thoughts-smart-city-rem-koolhaas.html.

follows from this, in which the larger part of production is of the derivative type and in which only a limited few (the troublemakers) are able to really innovate. This is not sustainable for the industrial complex, which, thus, has perfectly learned how to deal with conflict, using all sorts of techniques: from violence; to financial pressure; to media exposure and display; to cooptation. In a few words: encoding, recruiting, aestheticizing.

Subversive action has already been internalized by the market, under the form of instruments for the creation of value, to increase sales and for marketing. This is clear, for example, when we consider the linguistic (and, thus, perceptive and operative) transformation of the word "hacker."

In his "Preface to Transgression" (on "Bataille: a Critical Reader", by Bolling and Wilson), Foucault (1991) explains how transgressions force limits, boundaries, and norms to recognize themselves, requiring them to deal with their imminent disappearance.

Transgression creates a space and innovates.

Grosz (2001) defines this process as spatial excess, a new dimension which is able to go beyond preconceptions, prejudices, and worries about utility, "beyond the relevance for the present, looking toward the future." The revelation and discovery of this excess depends on the possibility for transgression. Excess is in the "problematic", which is full of potential. The clandestine, the unacknowledged, the unofficial find their survival—beyond crime—in the transgression of social norms and limits, those same limits which have excluded them in the first place. They recycle trash, appropriate spaces, invent communication channels, create styles, fashions, and trends. They do not cross borders: They move on them. By moving them, they innovate.

14.2 The Past of the Future

De Certeau, Lefebvre, Maturana, Bateson, Bhabha, and other show how this system is a cybernetic system of the second order. Citizens continuously reprogram their space, appropriating, hybridizing, creating relations, reactions, and transformation in the system. Systematic transgression creates innovation. Using a term from Massimo Canevacci it is the methodological indiscipline, a composition of the polyphonic and undisciplined strides of myriads of uncoordinated individualities, actuating their own style of spatial re-appropriation (both physical and digital), continuously creating conflict, transgression, and movement along and across boundaries and interstices.

The industrial complex has already reacted to this scenario, trying to resolve Enzensberger's paradox intervening on languages and imagination, by encoding the roles of troublemakers.

Of course, this type of path is present across the whole world: cultural institutes, "factories" for digital arts, workshops for creativity flourish everywhere. Linguistic metaphors are in plain view. Spaces and events are created (co-working spaces,

incubators, hackathons). Creatives are co-opted (makers and hackers). They are transformed in precarious research laboratories (startups, incubators, fabrication laboratories). Value and scalability/replicability are created (acceleration). The few good ideas are taken and sold (exits), generating profit.
This model, which is potentially virtuous, has a number of disadvantages, most of them at the level of the social and political discussion. On the one hand, it quickly comes back to the paradox of the creative industry: by encoding, conflict and transgression are integrated and, thus, unable to innovate. On the other hand, it creates precarity, and dumps business risks onto a variety of different fragile subjects.

On top of that, by enacting these processes, a few large operators become able to promote the social imagination formats which are useful for the cultural infiltration of their business: Innovation becomes the chase for a single form of future, instead of opening up to the opportunity of a plurality of possible futures.

Technology's role, in all of this, is banalized, reducing the complexity of the perception of how it could be possible to find solutions to the planet's major problems. Technology becomes a fetish which can be a per se guarantee for solution for energetic, environmental, and social problems. It becomes perceived how small groups of people, by coming together during a 24-h hackathon, can produce an app or a website to confront with large issues which are political, social, and cultural, not certainly technical. This obviously is reductionist and simplistic.

In all of this, major environmental, energetic, societal, and financial issues loom over the earth, creating the told and untold anxieties and pressures for transformation to happen, driving anything from the rise of crises and wars, to sharing economies and technological innovation, to anything that sits in between, together with the mutation of practically every element of our daily experience.

14.3 The Present–Future

How can we preserve the possibility for conflict and transgression, to maintain all of their positive effects on the world—through generation of diversity, plurality, and radical innovation—starting from the possibility for critical visions, and the consequent possibilistic opening to the perception of a multiplicity of futures and imaginaries?

To look for a possible answer, it can be useful to adopt the metaphor of the garden.

As we have seen in previous chapters, in his Moving Garden, Gilles Clément explores the possibility for a new type of garden, emergent, mobile and in perennial mutation, which lives in friches, the abandoned, uncultivated lands, those which history denounces as the loss of power of man over nature.

"What if we lay a different gaze on them? Could they not be the new blank pages which we need?" (Clément 2005).

On the one hand, historically, form—controlled form—was considered powerful in protecting us from the diabolical residues of the unknown.

On the other hand, "*friches* have nothing to do with dying and decay. In their beds species abandon themselves to invention. Walking in *friches* is a continuous process of self-interrogation. [...] Could this great power of reclaiming and conquering space not be placed at the service of the garden? and of which garden?" (Clément 2005).

The Third Landscape is a moving challenge, with mutating borders and boundaries, in a state of perennial conversation. It is the weeds which grow in between bricks and train tracks. It is the natural space of our cities which has not yet been encoded.

In our cities, the largest part of biodiversity is found in the Third Landscape. It is an interconnective tissue, composed by residual spaces, which resist government and form. In this, it is transgressive. It is a multiplication of narratives. It is not a property, but a possibilistic space for the future. It is a mutating transgression, which operates in emergent ways through multiple points of view and intentionalities. It is a syncretic map which evolves together to the mutations of the residential, industrial, and commercial areas of the city.

It is the geography of the mutation of the city.

As we have seen in Chap. 3, a similar metaphor can be applied to Casagrande's concept of "ruins."

If, on the one hand, ruins represent a loss of power from human beings to nature, on the other hand, according to different aesthetics, they represent the life of the city, demonstrating its usages and non-usages: The action (and non-action) of human beings leads buildings into a different state, transforming them into ruins and, thus, producing the evidence of their and nature's history. Ruins, to all effect, constitute a shared, extremely usable and accessible source of knowledge and information. According to Casagrande, the Third Generation City is the ruin of the industrial city, and becomes real when it recognizes its own local knowledge, becoming part of nature.

It is possible to search for solutions in these types of metaphors. How?

14.4 The Future

A new aesthetic system is needed, a new sensibility, which allows to recognize the value (and, thus, to directly support) of the continuous stratification, in our cities (and, in general, in the environment) of the unconscious, of the transgressive, of the conflict, of the different, to attribute value to it, as a new construction material which is able to innovate and to preserve history and knowledge, and to transform spaces and processes.

From Bhabha's and Soja's Third Space, to Clément's Third Landscape, to Casagrande's Third Generation City, to Pistoletto's Third Paradise, to our Third Infoscape, alluding at its informational manifestations.

A new aesthetics, a new sensibility, a new imagination, corresponding to the possibility for institutions with a new form: ecosystemic; not only responsible for strategies, but also for the possibility for the emergence of tactics, transgressions, and conflicts.

Not only "normative actuators and certifiers," but also—and most of all—direct and responsible supporters of the environment in which transgression and conflict may take place, as a form of emergent, shared knowledge.

Going back to Clément's metaphor, together with the imagination for a new type of garden, we need a new conception of gardener:
"*it is hard to imagine which aspect these gardens will assume, in which existence is expected to assume no form. From my point of view, gardens of this kind should not be judged on account of their form, but, rather, on the basis of their capacity to generate and translate a certain joy of existence.*"

In our opinion, this is a perfect description of what Digital Urban Acupuncture tries to achieve: a state of participatory performance in which inhabitants of a territory discover new aesthetics and the desires which come with them, to understand the complex and ever-changing dynamics of their Relational Ecosystems in order to experience the joy of diversity and to actively contribute to it, together with their institutions and organizations, through punctual interventions. In a logic of interconnection, fluid evolution and pursuing the possibility for joyful, meaningful coexistence.

Gardeners of garden without a form.

References

Bataille G (1957) Eroticism. Editions de Minuit, Paris

Clément G (2005) Manifesto del terzo paesaggio. Quodlibet, Macerata

Enzensberger HM (1982) The industrialization of the mind. Critical essays. Continuum, New York

Foucault M, Bolling F, Wilson S (eds) (1991) Preface to transgression. Bataille: a critical reader. Wiley-Blackwell, London

Grosz E (2001) Architecture from the outside. MIT Press, Cambridge

Pine JB, Gilmore JH (1999) The experience economy: work is theater & every business a stage. Harvard Business School Press, Boston

MIX
Papier aus verantwortungsvollen Quellen
Paper from responsible sources
FSC® C105338

If you have any concerns about our products,
you can contact us on
ProductSafety@springernature.com

In case Publisher is established outside the EU,
the EU authorized representative is:
Springer Nature Customer Service Center GmbH
Europaplatz 3, 69115 Heidelberg, Germany

Printed by Libri Plureos GmbH
in Hamburg, Germany